BIOLOGY

AN ILLUSTRATED HISTORY OF LIFE SCIENCE

PONDERABLES™

100
DISCOVERIES
THAT CHANGED HISTORY
WHO DID WHAT WHEN

BIOLOGY

AN ILLUSTRATED HISTORY OF LIFE SCIENCE

Edited by Tom Jackson

Contributors: Richard Beatty • Leon Gray • Dr. Jen Green •
Tim Harris • Tom Jackson • Robert Snedden

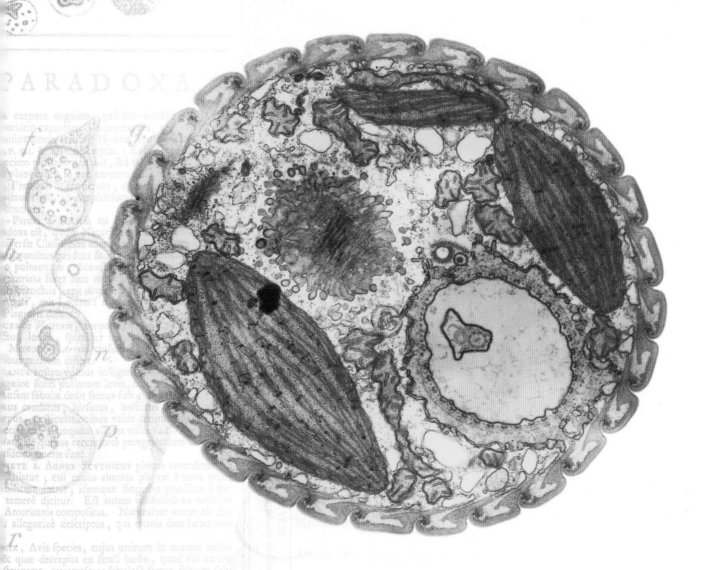

SHELTER HARBOR PRESS
NEW YORK

Contents

Introduction

BIOLOGY IS THE STUDY OF LIFE, AND THAT IS NO SMALL TASK. LIVING THINGS ARE THE MOST COMPLICATED OBJECTS IN THE UNIVERSE. A HUMBLE BACTERIUM HOSTS THOUSANDS OF CHEMICAL REACTIONS, ALL ORCHESTRATED TO KEEP THE CELL ALIVE, WHILE A BODY CONTAINS BILLIONS OF SIMILAR CELLS, ALL OF WHICH MUST WORK TOGETHER. A FOREST, A REEF, OR INDEED THE WHOLE PLANET, IS AN IMMEASURABLE COMMUNITY OF WILDLY VARYING LIFEFORMS THAT LIVE AND DIE DUE TO EACH OTHER'S INFLUENCE. HOW CAN WE MAKE SENSE OF IT ALL?

Biology often tests and breaks cultural boundaries. In the 17th century, William Harvey showed how the heart and blood supply worked by examining dead humans (and dying animals), which was a forbidden practice for earlier researchers.

Old ideas are used to make new ones.

The thoughts and deeds of great thinkers always make great stories, and here we have one hundred all together. Each story relates a ponderable, a weighty problem that became a discovery and changed the way we understand the world. Knowledge does not arrive fully formed. We have to work at it, taking it in turns to consider the evidence and offer our take on what is true and what is not.

LEARNING PROCESS

In hindsight, even the most cutting-edge ideas can turn out to be utterly wrong, but at the time, they are the best we have. Our civilization is built on knowledge—knowledge of plants, animals, and everything else around us—and that knowledge grows, step-by-step into an ever clearer picture of reality. Understanding life, what features all lifeforms share, and how they may

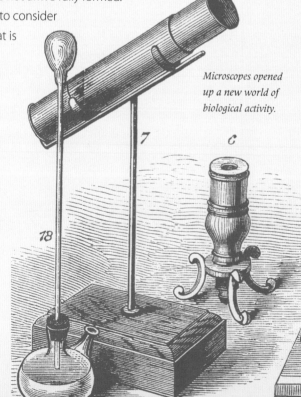

Microscopes opened up a new world of biological activity.

differ, is the best way to understand ourselves, where we fit, and how we can modify, enhance and, above all, protect our world.

LIVING WITH NATURE

When did humans become interested in nature? That question assumes that our species is not really part of nature, that it is something separate. Of course, the pages of this book, with its text and pictures, show that our species does something that other life does not, but we came from nature just as a flower, fish, or fungus did. In our earliest days, other lifeforms would have represented two things: Food or threat. About 15,000 years ago, something changed. Humans began sharing their habitats with other organisms. First came dogs, or tamed wolves, who benefited from life among humans, and helped us with protection and hunting—and perhaps with friendship, too. As the millennia passed, humans learned to keep animals for food and labor, and grew plants as crops, and only then, once nature had become a tool for us to use, did our ancestors begin to consider broader questions about how it worked and what great diversity it contained.

GOALS OF BIOLOGY

Early plant biology was tied up with medicine, as doctors learned to use different plants—herbs, bark, and sap—as therapies for disease (with mixed results). Animal biology probably began with hunting. Rulers sought out new, larger, and fiercer foes on which to test their killing skills. The end result was a growing list of known organisms, and people began to plot relationships and links between them. At last count, biologists have described 1.3 million species, and estimate there are 7 million more in need of a name.

This tiny Central American frog's bright colors warn predators that its skin is poisonous—don't touch!

Farmers also found that the features of an organism can be changed by controlling which individuals breed with each other. This process of inheritance would revolutionize our understanding of life by creating the fields of genetics and evolutionary biology. But, as well as seeking general rules that govern life, today's biologists look for specifics among the great diversity of organisms. In recent years, biological knowledge has become the focus of technology. What will that mean in the future for life as we know it?

A virus is DNA encased in proteins. Is it alive? Biologists cannot agree.

The contents of an organism's genes can be easily compared using profiling techniques, which also shows relationships, some expected, others less so.

A rainforest teems with life, but is a very fragile habitat. Just small changes will have large effects on the different wildlife that call it home.

Life on Earth

The diversity of life is hard to imagine, but this handy guide is a good place to start. Much of our focus is on mammals, animals that look and live like us, but we (the mammals) make up only about a quarter of one percent of the species on Earth.

Prokaryotes

The first organisms were single-celled bacteria and archaea. These lifeforms are known as prokaryotes, which means "before the nucleus" because, you've probably guessed already, they do not have a nucleus or any internal structures (organelles) in the cell.

BACTERIA

The most widespread form of life, bacteria are found in all habitats and even exist in rocks deep underground and the high atmosphere.

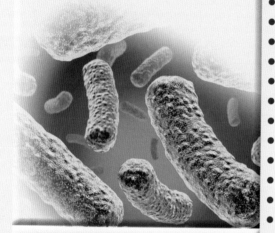

ARCHAEA

Often associated with extreme environments, such as hot springs and salty water, these organisms are possibly the earliest forms of life.

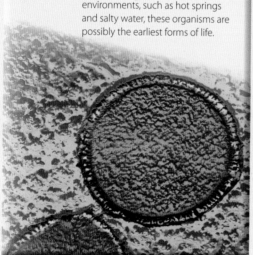

Eukaryotes

These organisms have cells containing nuclei, where the genetic material is stored. The cells also contain internal structures, called organelles, that perform specific functions. The eukaryotes include single-celled organisms as well as all multicellular species.

PROTISTS

This very large group of single-celled organisms include protozoa, amoebas, and algae. Protists of different kinds are thought to be the ancestors of all multicellular life.

PLANTS

Plants are responsible for turning Earth green. These organisms use a green chemical called chlorophyll to collect energy from sunlight, which is used to power a plant's body.

Seaweed
Appears in red, brown, and green forms.

Ferns
Spread using spores, not seeds.

Conifers
Produce seeds inside cones, but no fruits.

Mosses
Tiny plants without leaves, stems, or roots.

Flowering plants
Flowers are used for breeding, and seeds develop inside fruits.

FUNGI

Fungi are saprophytes, which means they consume whatever they grow on, releasing chemicals to digest it. Although they make mushrooms to spread seed-like spores, fungi are largely unseen organisms that live in soil. They are important for rotting down dead material.

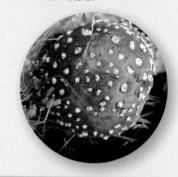

ANIMALS

Animals are active, mobile creatures, at least for some stage of their life cycle. They all must consume the nutrients required to fuel and build their bodies. This chart shows the main groups.

Sponges
Simple tube-shaped creatures that filter food from water.

Jellyfish
Circular animals that also include sea anemones and corals.

Worms
A collection of small animals with elongated, limbless bodies.

Mollusks
Mollusks mostly have shells, but also include squids and octopuses.

Crustaceans
Mostly aquatic animals with multiple limbs and appendages.

Insects
Mostly terrestrial six-legged animals with an external skeleton. Some fly with two or four wings.

Arachnids
Mostly eight-legged animals that include spiders, ticks, and scorpions.

Echinoderms
Including starfish and sea urchins, these animals have bony plates in the skin.

Fish
Swimming animals with several fins; they breathe with gills.

Chordates
All vertebrate animals, those with a spine and internal skeleton, belong to this group.

Mammals
These are hairy animals that give birth to their young and feed them on milk.

Reptiles
Scaly animals that include lizards, snakes, turtles, and crocodiles.

Amphibians
Includes frogs and salamanders; early life is mostly in water.

Birds
Winged vertebrates that lay eggs.

1 The Menagerie

THE HUMAN INTEREST IN OTHER LIVING THINGS BEGINS WITH FOOD. WE'VE BEEN TENDING crop plants and keeping livestock for around 13,000 years. However, our wonder at the beauty and diversity of living things probably began in a menagerie.

In the 2nd century BCE, Hannibal, a general from Carthage in North Africa, used a force of elephants to invade Roman Italy.

A menagerie is a private collection of animals, which is put on display to entertain and inspire the public—and boost the status of the owner. Menageries were most popular in the 18th century, when many European aristocrats competed by showing off the most exotic creatures they could find, and this heralded the foundation of the first zoological gardens, or zoos. However, the menagerie has a much longer history than that.

Animal city

An ancient Egyptian carving shows a hippopotamus giving birth, with a Nile crocodile ready to strike.

The oldest menagerie known is a relatively recent discovery. In 2009, excavators at Nekhen, an ancient city by the Nile in central Egypt, found more than a hundred animal skeletons. The remains had been buried in the same way that humans would have been, and the assumption is that they were members

ISHTAR GATE

Nebuchadnezzar II, the king of Babylon, is famous for building the Hanging Gardens of Babylon and for being a mighty warmonger, known as the "Destroyer of Nations." The lush Hanging Gardens were said to be growing on the side of an artificial mountain built to please his Persian wife, who disliked life in the arid desert. Nebuchadnezzar's upgrade to the city also included the Ishtar Gate, a portal to the inner city reserved for Babylonian nobility. The gate was covered in pictures of lifeforms, especially lions and flowers. The most notable animals were the dragon, which represented the god Marduk, the patron deity of Babylon (and a favorite of the king), and the aurochs, the wild ancestor of today's domestic cattle, which represented Adad, the rain god, who needed to be appeased to ward off famine.

of the royal menagerie. At the time of these burials, around 3500 BCE, Nekhen was the largest city in Egypt and the capital of the upper kingdom (which was to the south of the lower one). The ancient Greeks called the city Hierakonpolis, which means "the Hawk City," an allusion to its association with the bird-god Horus. The menagerie, which included hippopotamuses, baboons, elephants, and wild cats, was a symbol of the ruler's god-like power. On his death (who it was exactly is not known), the animals were sacrificed and laid to rest shrouded in fine cloth on beds made from reeds.

Royal collections

Later rulers kept menageries for less pious reasons, such as keeping game animals for the hunt. The Empress Tanki, who ruled China in the 12th century BCE, built a marble "House of Deer," the earliest record of a zoo in that country. In the 4th century BCE, Alexander the Great sent strange animals back to Greece as his armies invaded great swathes of Asia. As a child Alexander had been tutored by Aristotle, one of the greatest thinkers in history. Aristotle is famed for his ideas about logic, physics, and ethics. However, he is less remembered for being the founding force behind another science: Biology.

ASHUR-BEL-KALA

As the king of Assyria (in what is now Iraq and Syria) in the 11th century BCE, Ashur-bel-kala received frequent gifts from his powerful neighbors in Egypt. Many of the gifts were exotic animals collected from the African interior, including a large ape (probably a gorilla), a crocodile, and a "river man" (which might be a dugong, or sea cow), plus other "beasts of the Great Sea." The king built an enclosure for his animals near his palace in Assur, and sent envoys all over the world to find new creatures—some to show off and others to test his hunting skills.

2 Aristotle's Animals

ARISTOTLE IS WIDELY CONSIDERED TO HAVE BEEN THE FOREFATHER OF THE SCIENTIFIC STUDY OF LIFE.
Aristotle's works brought together biological knowledge into a consistent whole, and remained the authority on the living world for centuries.

Aristotle was the son of the Macedonian royal family's court physician. Like his father, he trained in medicine, later going to Athens to study philosophy with Plato, where his career as a great thinker reached the history books. Aristotle traveled widely and made detailed observations of living things, especially aquatic animals which he studied in the large lagoon on the island of Lesbos. He looked for order among the variety of life and tried to explain it in a series of six books. Other thinkers before him had speculated about the natural world but Aristotle was the first to combine theory with investigation and experiment. Aristotle's biology was far from flawless. For example, he believed the purpose of the brain was to keep the body cool and that thinking took place in the heart. He also believed in spontaneous generation, the idea that life could spring from non-living material.

As an older man, Aristotle returned to Macedonia to serve as a tutor to the king's son Alexander (destined to be Great). He instilled in his pupil a fascination for animals and nature—and perhaps conquering, too!

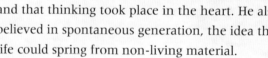

Aristotle was not aware of the rhinoceros, but his pupil Alexander the Great sent examples home from India. It is assumed that the legend of the unicorn is derived, at least in part, from this horned beast.

Living groups

Aristotle saw how plants and animals could be classified according to their physical make-up and by their habits. He split the animal world in two,

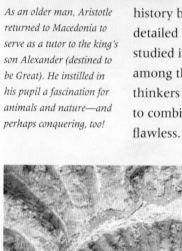

XENOPHANES

The philosopher-poet Xenophanes spent time in various parts of the Greek world during the late 6th and early 5th centuries BCE. He apparently lived to a ripe old age for the time—by his own reckoning he "tossed about the Greek land" for 67 years from the age of 25. Xenophanes believed that the world had been formed from water and "primordial mud." He was the first person known to have formulated a theory of the Earth's history based on the discovery of fossils. The fossilized remains of marine creatures found inland, far from the sea, inspired him to suggest that there had been alternating periods of worldwide flood and drought.

dividing it into those which had blood and those which did not (or at least didn't have red blood). To a great extent, this division reflects our present crude division of animals into the vertebrates (animals with a backbone) and the invertebrates (those without).

Aristotle grouped animals with broadly similar characteristics into genera (singular, genus), groupings that are still used by biologists today, although not in quite so wide a sense as Aristotle used it.

Aristotle's five genera of red-blooded animals were: Four-legged animals that give birth to their young (mammals), birds, four-legged egg-laying animals (reptiles and amphibians), fish, and whales. Dolphins were in the final category. Aristotle correctly saw they were not fish but he did not see them as mammals, either. The word *dolphin* comes from the Greek for "womb fish."

The bloodless animals he classed as cephalopods, such as the octopus and cuttlefish; crustaceans; insects, which included spiders and all other creepy crawlies (many of which are not insects in the modern sense); and shelled animals, such as mollusks and starfish. (The mouth structure of the sea urchin is called Aristotle's lantern due to his original keenly observed description.) The final animal group was the "zoophytes" or "plant-animals." These included jellyfish and anemones, which, Aristotle reasoned, shared features with both plants and animals.

Order of things

Aristotle was fascinated by the zoophytes, which seemed to have a rather blurred position in the scheme of things. He came to view nature as a continuum, stretching from the lifeless rocks, through increasingly complex plants and animals, and culminating in the human race. It was this view that would give rise in later centuries to the idea of the Great Chain of Being, a rigid hierarchy of life.

This is probably Aristotle's last legacy with respect to biology, an overarching sense that persists to this day, that small organisms are primitive and have naturally developed towards larger, more advanced life. The idea that living things (including us) are driven to develop in a particular direction—toward a final goal—was one argued by Aristotle, and it is still an assumption of Western culture, but *not* part of biology.

GREAT CHAIN OF BEING

The Great Chain of Being was central to Western thought from the time of Aristotle until around 1800. It was founded on three principles: Every possible kind of life imaginable exists in the universe; each species differs from its closest relative by an almost imperceptible degree, so all lifeforms are finely graded from one to the next; and all species have a place on a great chain, or ladder, of being that extends upward from the lowest form of life to God Himself.

3 Theophrastus's Plants

ARISTOTLE'S BIOLOGY WAS MOSTLY FOCUSED ON ANIMALS, but his student Theophrastus was the founding figure of botany, or plant science.

Theophrastus was born on the Greek island of Lesbos around 372 BCE. He studied in Athens under Plato and later Aristotle, and, following Aristotle's death, became head of Aristotle's school, the Lyceum, in Athens. In his multi-volume text, *Enquiry into Plants*, Theophrastus set out to do for plants what Aristotle had done for animals and find a scheme for classifying them into different types. He divided them into trees, shrubs, "undershrubs," and plants. The ninth, and final, book of the *Enquiry* considers the medicinal properties of plants and their other uses. It was hugely influential on later scholars of medicinal plants.

In another work, *On the Causes of Plant Phenomena*, Theophrastus looked at plant physiology and considered different methods of cultivation. He had an awareness of the relationship between plants and their environment and how plants were adapted to different conditions such as moisture, temperature, and soil type, explaining that plants needed a "favorable place" if they were to thrive. Theophrastus died at the age of 85. He expressed a wish that he be buried in his garden, and left instructions for the garden's care following his death.

Theophrastus began the daunting task of applying specific names to plants, such as the maple, which was previously called several different names by various communities in ancient Greece.

CHINESE HERBALISM

The use of plants in medicine is a tradition that stretches back for thousands of years. There are ancient Chinese writings on the subject dating from around 3000 BCE, and medical herbalism is likely to have been practiced long before that. The mythical emperor Shennong ("Divine Farmer"), said to have been born in the 28th century BCE, was reputedly the first herbalist. His knowledge of medicinal and poisonous plants is said to form the basis for the study of Chinese herbalism. One of the discoveries Shennong is credited with is that of tea. He claimed that it was an antidote for 70 different kinds of poison!

An extract of Plants of the Southern Regions, *written by Ji Han in the 4th century CE.*

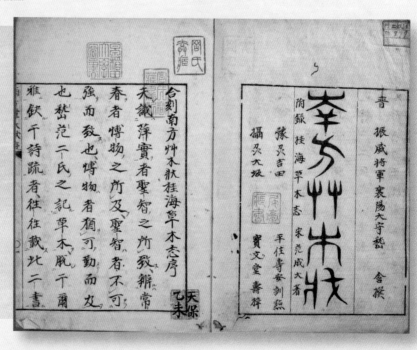

4 Panspermia

THE QUESTION OF HOW LIFE FIRST AROSE ON EARTH IS ONE TO WHICH THERE IS, AS YET, NO ANSWER. One idea given serious consideration was panspermia, which says that life arrived on Earth from outer space.

The earliest recorded proposal for panspermia (it means "seeds everywhere") came from the 5th-century-BCE Greek philosopher Anaxagoras, who asserted that the seeds of life are present everywhere in the Universe. Both plants and animals had their origins in panspermia, arriving on Earth via "meteors." This term was used to describe any unknown light in the sky. Most meteors of old would probably be identified as shooting stars today, and we know they are from specks of dust—or occasionally larger bodies—that hit Earth's atmosphere. However, in ancient Greece, meteors also included other atmospheric events, and that is why meteorology is the study of the weather, not meteors.

Revival of the concept

So the idea of panspermia is very old, and it has gone in and out of favor over the intervening years. It was addressed in a more scientific form in the 19th century by the Swedish chemists Jöns Jacob Berzelius and Svante Arrhenius, and the Scottish physicist Lord Kelvin. In the 1970s cosmologists Fred Hoyle and Chandra Wickramasinghe championed panspermia again, even suggesting that extraterrestrial lifeforms continue to enter the Earth's atmosphere, and were responsible for disease epidemics.

In 2009, eminent physicist Stephen Hawking gave his support to the idea, declaring that life could be spread from planet to planet on meteors. The idea that life came from space is not impossible; rock-eating bacteria could survive deep inside a meteor. But it does rather beg an answer to the question of where and how life originated. All panspermia does is push the answer to some unknown, far-flung part of the Universe, which is really no answer at all.

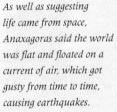

As well as suggesting life came from space, Anaxagoras said the world was flat and floated on a current of air, which got gusty from time to time, causing earthquakes.

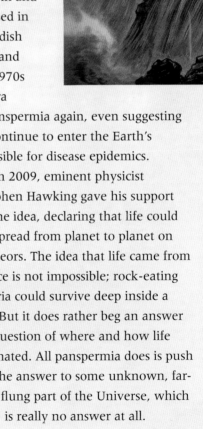

Anaxagoras said that the Sun and stars were fiery stones that were at a great distance from Earth. He took meteor showers, where shooting stars appear in clusters, as evidence of this.

5 Preformationism

PREFORMATIONISM WAS THE MAIN THEORY ON HOW PLANTS AND ANIMALS GREW FOR MOST OF HISTORY. It says an organism is a miniature, but fully formed, version of its mature form from the moment of its creation.

PYTHAGOREANS

The Greek scholar Pythagoras, who lived around 350 BCE, put forward one of the first theories to explain why it was that offspring resembled their parents. He believed that semen traveled through a man's body, gathering the essence of all its characteristics, such as eye color, skin tone, and musculature. The semen was like a condensed essence of the man himself. Basing his world view entirely on geometry, Pythagoras described the developing offspring as the "nature" provided by the father and the nurture from the mother working like two sides of a triangle to determine the third length—the child.

The earliest record of embryological research has been attributed to Hippocrates in the 3rd century BCE. Hippocrates believed that the embryo's development depended on it extracting moisture and breath from the mother. He was also one of the first to raise the idea of preformationism, as he suggested that organisms were fully formed in miniature inside a tiny packet or egg. The rival theory to preformation, epigenesis, was proposed by Aristotle, whose observations of developing chicken embryos showed that organisms develop gradually over time. Aristotle contended that each body part of the parent delivered even tinier packets to the embryo, which were then assembled to make a new body.

Hippocrates wrote about embryology in his books on obstetrics and gynecology.

The revival of preformationism

In the 15th century CE, the Dutch biologist Jan Swammerdam lent his support to a revival of preformationism after observing folded butterflies inside chrysalises. According to Swammerdam, adult butterflies were preformed inside caterpillars. He insisted that the various stages of the insects he observed, from egg to larva, pupa and adult, were nested one inside the other like a series of Russian dolls.

By the early 1700s, preformationism had become firmly established as a theory of embryological development with a variety of researchers, including noted microbiologist Antonie van Leeuwenhoek, claiming to make out a miniature man, called a homunculus, inside sex cells. These observers were divided between those who believed that the homunculi were to be found in eggs (ovists) and those who believed they were in sperm (spermists).

Accurate observations of embryological development were only made in the 19th century after greatly improved microscopes were built. Finally, preformationism was overthrown in favor of a better understanding of cell theory and then genetics.

An alchemist searches for the secrets of life, hoping to create a child by cooking up a homunculus in his workshop.

6 A Natural History

ALTHOUGH IT IS A CURIOUS MIX OF FACT, HEARSAY, AND SPECULATION, PLINY THE ELDER'S MASSIVE 37-VOLUME UNDERTAKING, *Historia Naturalis* (Natural History), was the world's first encyclopedia of science and the main source of information on animals in Europe for several centuries.

Roman scholar Gaius Plinius Cecilius Secundus, known as Pliny the Elder (c.23–79 CE), was born near what is now the town of Como, Italy. Educated in literature and the law, he also had a lengthy military career, reaching the rank of cavalry commander. He served under three emperors, acting as a legal advocate during the reign of Nero and serving on the imperial council for Vespasian and Titus. Somehow, he still found time to compose more than 75 books, the most famous of which, and the only one to survive, is his *Natural History*.

Information and misinformation

Over the course of *Natural History*'s 37 volumes Pliny explored such diverse topics as plant and animal biology, human physiology and anthropology, and geography and astronomy. Notwithstanding the fact that many of his suppositions and assertions would later prove to be false, this is still a valuable insight into the history and culture of Rome in the first century CE.

Zoology was covered in books VII to XI, and botany in books XII to XVII. Pliny offered detailed descriptions of the life and habits of the hippopotamus, among which was the assertion that when the animal grew too fat, it would deliberately back into a sharp stick to bleed itself back into good health. The work also includes detailed descriptions of unicorns and other creatures told of in myths. For example, he explains that the basilisk, a six-legged beast with the head of chicken, could kill you with a stare. Pliny regarded myths as true accounts, and simply assumed that the fierce, and perhaps magical, animals, were very rare or just difficult to catch.

PLINY'S DEATH

One of the best accounts we have of the eruption of Vesuvius and the destruction of Pompeii comes from Pliny's nephew, known as Pliny the Younger. According to his account it is clear that the disaster caught everyone unawares. His uncle was the senior military officer in the area at the time. Apparently, he took a small vessel ashore near Herculaneum to investigate what was happening and succumbed to asphyxiation from the fumes of the eruption.

Pliny declares in the preface to his book that it contains 20,000 facts. Almost as many as this book!

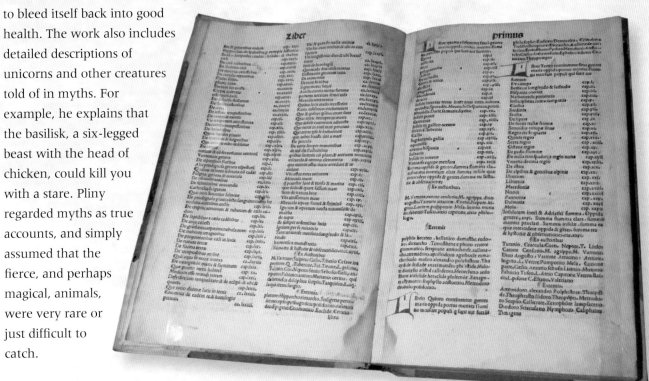

7 A Life Force

THE SIX BOOKS OF THE EPIC POEM *DE RERUM NATURA* (ON THE NATURE OF THINGS) set out to explain no less than the origin, structure, and destiny of the Universe and the emergence and evolution of life.

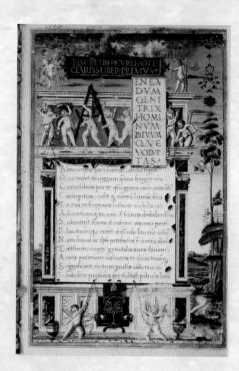

We know very little about Titus Lucretius Carus, the author of *De Rerum Natura*, other than that he lived around the 1st century BCE and followed the teachings of the philosopher Epicurus (who said that the secret of a good life was to seek enjoyment). The writing of *De Rerum* also reveals Lucretius to have been a serious student of philosophy and science, a keen observer of nature with a deep knowledge of crops and horticulture, and with an appreciation of landscape and the changing seasons.

A copy of De Rerum Natura *from 1483 produced for Pope Sixtus IV.*

GALEN

Claudius Galen (130–210 CE) was born in the city of Pergamon, in what is now Turkey. He studied medicine, becoming chief physician at the gladiator school, where he became an expert in treating wounds. He moved to Rome in the early 160s where he became physician to the emperor Marcus Aurelius and his successors, Commodus (the bad guy in the *Gladiator* movie) and Septimius Severus. Galen was a pioneer of medical experiments and sought to further his knowledge through the dissection of animals (cutting up humans was illegal). Galen said that pneuma, the vital spirit, entered the body through the lungs, where it was then distributed by the blood.

One thing that becomes apparent is Lucretius's commitment to naturalism. Throughout *De Rerum Natura* he persistently rejects supernatural explanations of natural phenomena. The world according to Lucretius was not divinely created but the result of entirely natural, ongoing, random processes. However, he also described a life force under the control of the goddess Venus. This is at odds with his other naturalistic ideas about living things. However, the concept of a "vital force" gained traction among later thinkers, especially doctors. Life, it was said, arose from a supernatural supply of *pneuma*. This could be translated as "breath" but also means "spirit." Pneuma was what set life apart from non-living materials, and for the next 1,800 years it was assumed that the processes of life, as conveyed by pneuma, set it apart from other natural phenomena.

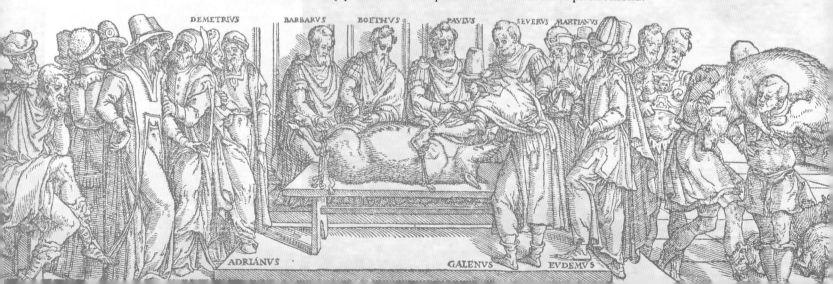

DEMETRIVS   BARBARVS   BOETHVS   PAVLVS   SEVERVS MARTIANVS

ADRIANVS   GALENVS   EVDEMVS

8 Herbals

SO VALUABLE WERE PLANTS AS A SOURCE OF MEDICINES, to say nothing of their various culinary uses, it is no surprise that most of the early works on plant science were "herbals", or guides to identifying plants with useful properties.

Dioscorides, a 1st-century-CE Roman army physician, was the author of the *Materia Medica*, a comprehensive account of almost 600 medicinal plants. For 1,500 years, it was an essential guide: A plant had to be listed in the *Materia Medica* to be considered of genuine value. Little progress was made in the study of plants during the Middle Ages, people tending to rely on the earlier work of people like Theophrastus and Dioscorides. The best-known naturalist of the period was Albertus Magnus, who worked in the 12th century. He was known to his contemporaries as Doctor Universalis because of the scope of his interests. Magnus's book *De Vegetabilis* gave first-hand accounts of a variety of plants and their uses. He was the first person known to identify certain plant characteristics, for example he separated monocots (such as grass) and dicots (such as beans) based on their stem structure. The invention of the printing press in the mid-15th century saw a surge in the publication of books on plants. These herbals came to rely less and less on traditional lore and more on observations of plants in their own right, rather than purely as sources of medicine.

An anatomical drawing from a 13th-century English herbal shows leaf and flower shape.

Kitab-i hasha'ish *(The Book of Herbs) was an 11th-century Arabic translation of the* Materia Medica.

9 The Bestiary

DURING THE MEDIEVAL PERIOD, THERE WAS A COMMONLY HELD BELIEF, PARTICULARLY IN THE CHRISTIAN WORLD OF EUROPE, that the natural world was provided by God as a source of instruction on how people should live their lives. This information was told as stories about creatures in a great compendium of beasts, or a bestiary.

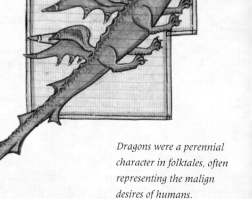

Dragons were a perennial character in folktales, often representing the malign desires of humans.

The first known bestiary text to set out these teachings was the *Physiologus*, written in Greek in Alexandria around the 2nd or 3rd century CE by an unknown author. Explicitly based on Christian beliefs, this collection of animal lore first describes some aspect of animal behavior and then instructs the reader as to how it should be interpreted as a way of learning Christian beliefs and practices through observation of the natural world. The *Physiologus* was hugely popular and translated into most of the major languages of Europe and western Asia. It is the primary source for Medieval bestiaries.

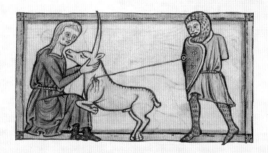

The only way to catch a unicorn was to lure it toward you using a fair maiden or two.

Etymologiae

Around the 7th century, Isidore of Seville wrote his *Etymologiae*, an encyclopedia that was an attempt to set down all that was then known about a huge range of knowledge, including military science, astronomy, theology, agriculture, and zoology. There is very little that is original in the *Etymologiae*, as much of it was derived from the books of earlier authors, such as Aristotle and Pliny the Elder. As the book's name suggests, Isidore sought to connect the names of things (their etymology) to their natures and characteristics. Some of these connections were more imaginative than scientific. For example, his suggestion that the Latin name for dormice, *glis*, refers to them growing fat and swelling (*gliscere*, in Latin) before they hibernate in winter.

A wyvern, a two-legged dragon, introduces itself to a startled elephant.

As can be seen in this picture of a lion, wolf, and cow, illustrations were not anatomically correct.

The birth of the bestiary

The book type known as the bestiary was born when the *Etymologiae* and similar texts were combined with the *Physiologus*. Isidore had simply set down whatever he gleaned from the earlier writers about the habits of animals, real and imagined, but without any additional moralizing tales. The compilers of the bestiaries quoted Isidore and tagged on the appropriate lesson to be learned.

FOUR BEINGS

Born in Switzerland in 1493, Paracelsus was the most influential thinker of his day. He thought the world was constructed from four elements: Earth, water, fire, and air. He also proposed that each element had its own being, equal but different to humans, and who occasionally strayed into our world. Gnomes lived in the earth and sometimes attacked miners. Nymphs lived in water and appeared nude on the shore. Salamanders were wiry fire people, who lived in volcanoes but did not get burned. Finally, sylphs were giants that floated in the air among forest trees. When they landed, they made the ground shake, causing earthquakes.

White birds were said to cure the sick, by taking the disease and flying away with it.

The bestiary, or book of beasts, had much in common with the *Physiologus*, continuing with the tradition of using the natural world as a teaching tool, but, importantly, it also took steps toward becoming a description of the diversity of the natural world in its own right.

Bestiary manuscripts were usually illustrated, although these Medieval animal illustrations are often far from realistic because it was highly unlikely that the artists had any first-hand knowledge of the creatures they portrayed. Instead they relied on travelers' tales and hearsay. Crocodiles were frequently portrayed as dog-like beasts in Medieval bestiaries, for example. The illustrations were really intended to serve as a "visual language" for a largely illiterate public. Many of the stories would be well known already, and the image of the creature in question would simply trigger recall of it.

Mythical creatures

As well as having unrealistic depictions of real animals, the bestiaries were populated by creatures that were not real at all. Unicorns, the griffin, and the basilisk were portrayed alongside the ostrich (which was given hooves) and the whale (shown as a big scaly fish). We don't know whether the people of the time actually believed in these creatures, though no doubt many did. The doubters probably didn't believe in the hippopotamus either.

Basilisks were thought to kill with their bad breath.

By the 13th and 14th centuries, a series of encyclopedias appeared which, while they contained some bestiary material, dispensed with the use of allegory.

Owing to the influence of people such as Albertus Magnus and Saint Francis of Assisi, interest was growing in studying

Satyrs, mischievous spirits from Greek myths, became the inspiration for the European image of the devil.

nature in its own right. Likewise, world travelers, such as Marco Polo, journeyed to the places where exotic creatures were supposed to dwell and failed to find them. As the Renaissance dawned in Europe, there was a resurgence in a more scientific approach to the study of nature that set out to discover fact rather than revel in fantasy.

10 Spontaneous Generation

IN THE ANCIENT WORLD THERE WAS A FIRM BELIEF THAT LIVING CREATURES arose from non-living material. This "spontaneous generation" appeared to take place mostly in decaying matter, such as when maggots appeared in rotting food.

Meet the ancestors? Believers in spontaneous generation proposed that simple animals emerged from rotting matter and then gradually evolved into more complex organisms— like you.

Jan Baptist van Helmont was a pioneer of pneumatic chemistry. He showed that burning wood gave out a stifling gas, which he called the "chaos of the forest," but which is better known today as carbon dioxide.

Aristotle accounted for spontaneous generation as resulting from a "vital heat" that was present in everything. This idea, as with so many of his, was widely accepted as fact for the next 2,000 years or so. The Medieval Church modified Aristotle's vital heat, pronouncing that spontaneous generation was carried out by angels acting through the agency of the sun.

Jan Baptist van Helmont (1580–1644) was a skilled chemist who was one of the first to isolate gases from air. He believed that knowledge of the natural world was best acquired through experiment. Yet he also published a recipe for producing mice! This involved placing rags in an open pot with a few grains of wheat and waiting 21 days for the mice to appear. There will, he assured his readers, be adult males and females present, capable of mating and reproducing more mice. The idea of spontaneous generation was a remarkably persistent one; even such eminent thinkers as René Descartes and Isaac Newton subscribed to the theory. In 1688, the Italian naturalist Francesco Redi demonstrated that maggots grew from fly eggs. If flies were kept away from meat, the maggots would not appear. There was nothing spontaneous about it.

A family get-together. Do geese and barnacles share a natural link?

GEESE AND BARNACLES

People in the Middle Ages were puzzled by a species of goose. It appeared each winter in Ireland and Scotland, but where did it nest? The birds' coloring matched that of a type of barnacle found on the seashore. That must be it: The "barnacle geese" did not have nests, but arose from "goose barnacles." It was an odd notion—Albertus Magnus dismissed it as "altogether absurd"—but the idea stuck long enough for both species to still go by those old names. If you were wondering, the barnacle goose migrates in spring to nest in the Arctic.

11 Finding Homologues

THE IDEA OF HOMOLOGY FIRST APPEARED IN MATHEMATICS, where it denoted an equivalence between sets of numbers. Its use in biology to map out shared characteristics between living things had its origins in the work of Pierre Belon. Homologous features in animals suggest that they share a past.

Pierre Belon (1517–64) was one of the first naturalists to make anatomical comparisons between different forms of life. He was born in Le Mans, France, and studied medicine in Paris. One of the main participants in the 16th-century revival of natural history studies, Belon is mainly known for his *L'Histoire de la Nature des Oyseaux* (The Natural History of Birds). Published in 1555, it described 200 species and pointed out similarities of form and function between the skeletons of birds and humans, so was a first step toward the discipline of comparative anatomy. Belon introduced the idea that different animals—or plants—develop in different ways using the same initial anatomical structures. Later biologists, such as Charles Darwin, suggested that organisms that share many anatomical structures were closely related.

In the early 19th century, Georges Cuvier suggested that the range of homologous structures observed in related animals arose because of their interactions with the environment. Étienne Geoffroy Saint-Hilaire proposed an alternative view: Animal bodies were all modified from a single unified plan. In a prelude to Darwin's theory of evolution, the British anatomist Richard Owen proposed in 1849 that skeletal homologies suggested that four-legged land animals developed from fish.

Pierre Belon found that human and bird skeletons were made of the same set of bones, that were modified to provide different functions.

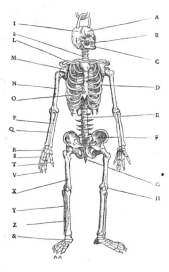

LIVRE I. DE LA NATVRE

Portraict de l'amas des os humains , mis en comparaison de l'anatomie de ceux des oyseaux, faisant que les lettres d'icelle se raporteront à ceste cy, pour faire apparoistre combien l'affinité est grande des vns aux autres.

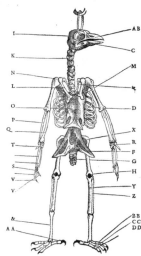

DES OYSEAVX, PAR P. BELON.   41

La comparaison du susdit portraict des os humains **monstre com-** bien cestuy cy qui est d'vn oyseau, en est prochain.

Portraict des os de l'oyseau.

12 Zoology

CONRAD GESSNER'S *HISTORIA ANIMALIUM* (HISTORY OF ANIMALS), published in 1558, was the first attempt to compile all the information then known about the world's animals. It is a landmark work on the road to modern zoology.

Conrad Gessner was a true polymath, with a reputation not only as a naturalist but also as a scholar of classical languages, a botanist, and a physician.

Gessner was born in Zurich, Switzerland. After receiving his doctoral degree in 1541, Gessner spent the rest of his life practicing medicine in Zurich. However, in his spare time he indulged his hobby of observing wildlife and investigating their anatomy. Taking his cue from Aristotle's classification of genera, Gessner divided the animal kingdom into four main categories: Quadrupeds, amphibians, birds, and fish (and other aquatic animals), devoting a volume of the *Historia* to each. A fifth volume, on snakes and scorpions (which included spiders and insects), was published in 1587, after Gessner's death.

Gessner was not only a gifted scholar, he was also a talented illustrator. This colored woodcut from Historia Animalium *is of the bald ibis.*

Ancient to modern

Rather than rely on the anecdotal work of earlier writers like Pliny and Aristotle for his information, Gessner took a more empirical approach. He focused his research on accurate descriptions, as far as possible based on his own observations and dissections. "The scholar," said Gessner, "should … observe, dissect, describe, and illustrate the animals himself." This isn't to say that Gessner ignored older sources completely: As well as acknowledging the input of Aristotle and others, he quoted extensively from the Old Testament and the Medieval bestiaries, even including creatures of legend such as the unicorn, phoenix, and sea serpent. By adding to past knowledge by including his own original work Gessner was, in many ways, building a bridge between the chroniclers of the ancient and Medieval worlds and a more modern approach to science.

Illustrated details

The *Historia Animalium* was lavishly illustrated by some 1,000 hand-colored woodcuts depicting the animals, many of them executed by Gessner himself. But *Historia Animalium* was far from being a simple depiction of animals in their natural settings. Gessner also discussed modern notions of habitat, behavior, and physiology, as well as dealing with the part played by animals in literature and art. He even found room to discuss their preparation as food and medicine.

v I s, cuius hic effigies habetur, à noftris nominatur uulgo ein Waldrapp,
fyluaticus, quòd locis fyluofis, montanis & defertis degere foleat: ubi in rupi

13 Water and Life

FOR AS LONG AS PEOPLE HAVE BEEN ABLE TO THINK, IT HAS BEEN ABSOLUTELY CLEAR THAT WATER IS ESSENTIAL FOR LIFE. For the philosophers of ancient Greece and elsewhere, it was one of four elements—the others were fire, earth, and air—that formed the world. A simple experiment in the 17th century put this idea on a scientific footing. Was water the source of all life?

As well as proving the significance of water in life processes, Jan van Helmont, who was an alchemist, also found the first evidence of carbon dioxide gas. He identified this gas in the smoke of burning wood, so called it "chaos of the woodland." But he made no link between carbon dioxide and life.

Aristotle had promoted the view that plants absorbed their food from the soil. This idea remained unchallenged until the 17th century. The Flemish scientist and physician Jan Baptist van Helmont was a disciple of Paracelsus and shared his alchemical belief that only air and water were fundamental elements, but not fire or earth. In what has become one of the best-known experiments of early biology, van Helmont set about disproving Aristotle's assertion.

MONISM

Monism is the belief that ultimately all of the stuff in the Universe arises from one fundamental material. It was a view widely held by the Greek philosophers. For Thales, a founding figure in Western thought, that material was water. As water is the only substance that can be readily observed as solid, liquid, and vapor and is so obviously vital for life, Thales's conclusion was perhaps not so surprising.

THALES.

Roots of science

Van Helmont placed a 2.27-kilogram (5-lb) willow tree in an earthenware pot containing 90.7 kg (200 lb) of soil. Over a five-year period, van Helmont tended the willow, giving it nothing but rainwater or distilled water. At the end of the time he weighed the tree again, discovering that it was now about 77 kg (169 lb); he also reweighed the soil, discovering that it was only a mere 56.7 grams (2 oz) or so lighter than five years before.

From this observation, van Helmont concluded that some 74 kilograms (164 lb) of "wood, barks, and roots," not even including the leaves the tree shed every autumn, "arose out of water only." He accounted for the lost 60 or so grams of soil as being down to experimental error.

Obviously, van Helmont could not have known that the soil provides minerals for making plant proteins and other biomolecules. And he knew nothing of photosynthesis, the process by which a plant harnesses the energy of sunlight to generate new plant tissue from carbon dioxide in the air, taking water from soil. However, he had uncovered one piece of the puzzle. It would be another 200 years before the pieces began to fall into place.

14 Blood Supply

IN 1628, WILLIAM HARVEY PUBLISHED ONE OF THE MOST IMPORTANT WORKS IN THE HISTORY OF SCIENCE. It demonstrated the circulation of blood through the body, and corrected an ancient mistake.

According to Galen, an influential Roman doctor from the 2nd century, blood was formed inside the liver from food. It then entered the veins and was carried to all parts of the body, to be transformed into new tissue or to provide nutrition for existing tissues. He thought that blood was constantly replenished as it was consumed by the body. This view persisted for the next 1,500 years. Harvey found this idea wanting, especially when he figured out that it would mean a person would be making 250 kg (550 lb) of blood every hour! In 1628, he published *An Anatomical Exercise on the Motion of the Heart and Blood in Living Beings*, which set out a demonstration of the circulation of the blood through the body.

Road to discovery

Until the 17th century, the body was believed to have two separate systems for carrying blood. One was composed of the veins, which transported nutrient-rich, purple blood from the liver to the rest of the body. The second system carried scarlet blood through the arteries, distributing life-giving pneuma, which was believed to be a vital force drawn in through the lungs.

William Harvey studied medicine at the University of Padua in Italy, learning from the scientist and surgeon Hieronymus Fabricius. Fabricius had made the discovery that the veins in the human body had one-way valves, but could not work out why this was so. It was Harvey who would explain the role the valves played in the movement of blood through the body.

After obtaining his medical degree, Harvey returned to England in 1602. He was a great success as a doctor, eventually becoming court physician for both James I and Charles I. As well as carrying out his medical practice, Harvey also continued his research. He was a dedicated experimenter and carried out dissections on a number of different species of animals to study their hearts and blood vessels.

The one-way valves in the veins, Harvey reasoned, meant that blood in those vessels could only travel toward the heart, not away from it. It was an important step toward

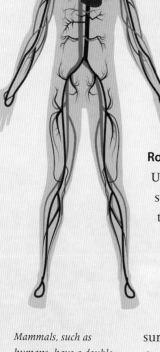

Mammals, such as humans, have a double circulatory system. A first loop of vessels takes blood from the heart via arteries (red) and then back again via veins (blue). The second, smaller loop carries blood through the lungs in order to exchange gases— oxygen comes in while carbon dioxide goes out.

SPIRITS

Galen taught the idea of pneuma as the fundamental principle of life. Pneuma physicon was located in the brain, and was involved with senses and movement; pneuma zoticon was in the heart, regulating body temperature; and another pneuma physicon had its home in the liver, controlling nutrition and metabolism.

Early ideas of neurology proposed that spirits flowed through the spaces in the brain.

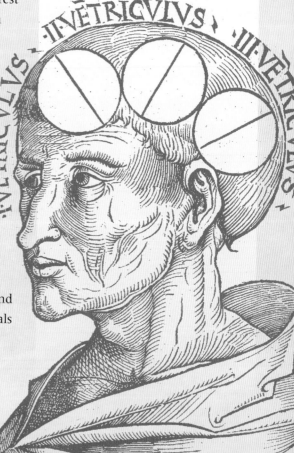

his understanding of blood circulation. He demonstrated that if he tied off an artery, the side nearest the heart expanded as it filled with blood. If he did the same thing to a vein, the side away from the heart bulged.

He also calculated that, over the course of a day, the amount of blood passing through the heart exceeded the body's daily intake of food by weight. He estimated that in one hour the heart pumped a quantity of blood that was equivalent to three times a person's weight. It seemed to Harvey that it was inconceivable that the blood was being replenished. It had to be the same blood moving in a constant flow through the body from heart to arteries to veins and back to the heart.

Harvey's discovery was based on scrupulously careful experiments and dissections of animals and people. Here he is studying the body of Thomas Parr, a man who had claimed to be 152 years old before he died. Harvey's conclusion was that Parr was no more than 70.

When Harvey published his results, he was met by ridicule and scorn for having dared to question the great Galen, whose views had been accepted by doctors for so long. But by the end of his life Harvey saw his ideas gain wide acceptance. The only gap in the argument was explaining how blood got from the arteries to the veins. Harvey rightly assumed that the connection was via vessels that were simply too fine to see. He was proved correct by Marcello Malpighi, who made use of the recently invented microscope to identify capillaries just four years after Harvey's death.

UNDERSTANDING MUSCLES

Giovanni Alfonso Borelli taught in a private anatomical laboratory at his home where his students included Marcello Malpighi. Borelli was especially interested in the structure and function of muscles and was one of the founders of iatromechanics, which sought to explain physiology in terms of mathematics and physics. His book *De Motu Animalium* was one of the first thorough studies of muscle physiology.

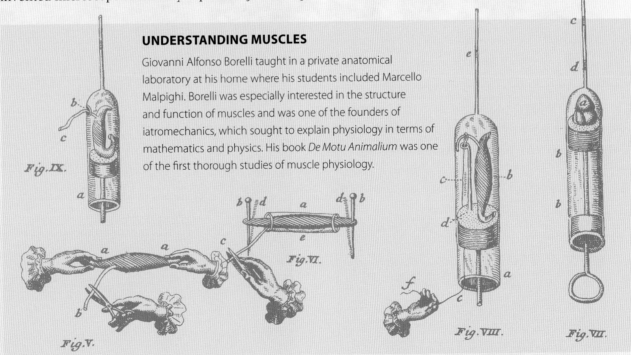

15 Metabolism

METABOLISM IS THE SUM TOTAL OF ALL THE CHEMICAL REACTIONS NECESSARY TO SUSTAIN LIFE. The term comes from the Greek word *metabole*, meaning "change." The study of metabolism began 400 years ago with a long, and smelly, experiment by Santorio.

Santorio settles down on his weighing chair to enjoy a meal. He is reputed to have made 32,000 measurements—that's three a day for 30 years.

Santorio Santorio—remembered by history by just one of his names (his first, but it makes little difference)—became a doctor at the age of 21. He practiced in Venice, where he socialized with the great physicist Galileo Galilei. Santorio's most famous experiment involved the study of body weight. He designed a movable platform suspended from an arm of an enormous balance. Sitting on the platform, Santorio would take his meals and record the change in his weight. If he remained on the platform he could see that his weight gradually declined again. He weighed his solid and liquid waste and found that this did not account for the entirety of the weight lost. After 30 years of continuous experiment, Santorio confirmed that the sum total of his visible excreta was less than the amount of substance he had ingested. He determined that the difference was made up through what he called "insensible perspiration."

Building and breaking

Insensible, or undetectable, perspiration had originally been proposed by Galen, but Santorio was the first to attempt to measure it. His book on the subject, *Concerning Static Medicine*, brought him fame and was the first systematic study of metabolism. Santorio's experiment shows that his food was providing the body with energy—although he was not equipped with that modern understanding. The energy was used to convert the content of his food into new body tissues and drive life processes. In the 20th century, biochemists began to show that metabolism is a cascade of reactions, with several thousand happening in a controlled manner within each cell.

There are two broad outcomes: Catabolism breaks up material, perhaps to release stored energy or to remove it from the cell. Anabolism builds up new material and is used in growth and maintenance.

Santorio attempted to explain the workings of the body on purely mechanical grounds. He argued that the body was like a clock, its operation determined by the shapes and positions of its interlocking parts.

THERMOSCOPE

Santorio is thought to be the inventor of the thermoscope, although some say Galileo did it. The thermoscope was a simple device that showed changes in temperature. Water was trapped inside a glass tube, and it rose and fell as it contracted and expanded with fluctuations of temperature. We do know that Santorio was the first to apply a numerical scale to the thermoscope, thus making it a more useful instrument.

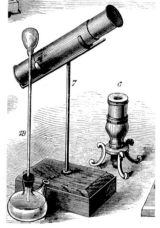

16 Embryos

EMBRYOLOGY IS THE BRANCH OF BIOLOGY THAT STUDIES THE EARLY DEVELOPMENT OF AN ORGANISM. Until the 18th century, the prevailing view was that the embryo was a fully formed organism in miniature. This idea was debunked by a scientific approach.

Also known as Girolamo Fabrizio, Hieronymus Fabricius wrote 16 books on anatomy and development.

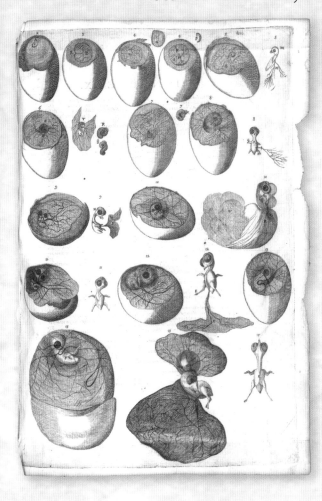

A drawing from a 1621 book by Fabricius shows the many stages of the development of a chick inside an egg.

In humans, the term embryo specifically refers to the ball of dividing cells from the moment they become implanted in the wall of the uterus until the end of the eighth week after conception. After that the growing baby, now with discernible features, is termed a fetus.

There are a few contenders for the title of "father of embryology," among them Marcello Malpighi, Caspar Friedrich Wolff, Aristotle, and even Leonardo da Vinci. Another with a claim to the title is Hieronymus Fabricius. Fabricius studied at the University of Padua under the great anatomist Gabriel Fallopius (also known as Gabriello Fallopio), and succeeded him as the university's professor of surgery and anatomy in 1562. It was while in this post that he taught William Harvey, discoverer of the circulation of the blood, his description of the valves of the veins, pointing Harvey in the right direction.

Internal development

It was the publication of Fabricius's *De Formato Foetu* (On the Formation of the Fetus) in 1600 that set the scene for the study of comparative embryology. In it he set out his observations of fetal development in several different species, including humans, and provided the first detailed descriptions of the structure of the placenta. Fabricius also inspired Harvey to study embryology. Examining a herd of deer that had mated, Harvey found no evidence of a developing embryo in the uterus until six or seven weeks afterwards and was convinced that the actual conception had taken place elsewhere in the body. It would be another century before the path of the fertilized egg through the Fallopian tubes to the uterus would be determined.

GABRIEL FALLOPIUS

Fallopius was an innovative anatomist, known for his talent for dissection. His main interest was in the anatomy of the head, but it was his investigations into the structure of human reproductive organs that won him his place in the medical textbooks. His description of the pair of fine channels, later shown to be the conduits for fertilized eggs to travel from the ovaries to the uterus, ensured that they now bear his name: Fallopian tubes.

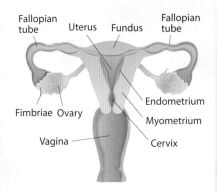

17 The Cell

THE CELL IS THE BUILDING BLOCK OF LIFE. Every living thing is made from at least one cell. Their discoverer was one of the most overlooked men of science.

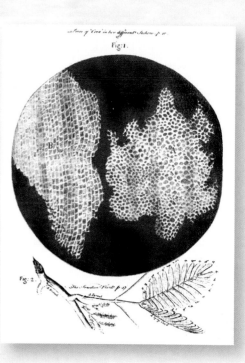

Hooke saw structures in a sliver of cork viewed under a microscope. He named these spaces "cells."

HOOKE'S MICROSCOPE

Hooke's microscopes were made by London instrument maker Christopher Cock, but Hooke himself was very much involved in their design. The microscope had separate draw tubes for focusing, and a ball and socket joint to direct the point of view. The lenses used in Hooke's microscope unfortunately were of poor quality. Hooke's attempts to correct this resulted in very dark images, so he designed a method of concentrating light on his specimens by shining the light from an oil lamp through a water-filled glass flask to produce an intense illumination.

A replica of Hooke's compound microscope and lighting system.

Robert Hooke was a scientific genius, making important contributions across a broad range of disciplines, not only in biology, but also in chemistry, physics, astronomy, geology, and architecture. Yet surprisingly, we have no idea what this great scientist looked like. No portrait of Hooke survives.

While a student at Oxford, Hooke impressed the eminent science professors with his skills at designing experiments and apparatus, and became an assistant to the chemist Robert Boyle. In 1662 Hooke was named Curator of Experiments for the newly formed Royal Society of London, with responsibility for demonstrating experiments at the Society's meetings.

Hooke earned his indelible place in the history of biology with his book *Micrographia*. Published in 1665, it recorded the observations he made with a compound microscope, an instrument he devised himself. With it he examined bird feathers, hair, insects, and other organisms. The book was illustrated with exquisite drawings, showing that Hooke was not only a first-rate scientist, but also a talented artist.

The observation for which he was most famous was his study of thin slices of cork. He saw that they were "perforated and porous, much like a honey-comb." He called these pores "cells," because they reminded him of monks' cells, or rooms, in a monastery. What Hooke saw were the cell walls of the cork tissue. He also reported seeing similar structures in other plants. The cells Hooke observed were empty—he thought they might be involved in transporting fluids through living plants. He appears not to have realized that the structures he had discovered were the basic units of living organisms. In 1678, after Antonie van Leeuwenhoek had reported his discovery of "little animals" to the Royal Society, Hooke was asked to confirm the Dutch scientist's findings. He successfully did so, thus helping to win acceptance of Leeuwenhoek's discoveries of what we now know to be single-celled protozoans and possibly bacteria.

18 Paleontology

PALEONTOLOGY IS THE STUDY OF THE HISTORY OF LIFE ON EARTH. IT IS FOUNDED ON AN EXAMINATION OF FOSSILS, the remains of once-living things that have been replaced by minerals or impressions of organisms preserved in rock.

A number of proposals had been put forward to explain the existence of fossils. One theory, dating back as far as Aristotle, was that fossils formed and grew within the Earth as a result of a "shaping force," which created stones that resembled living things. Leonardo da Vinci had no time for such an idea, declaring that "such an opinion cannot exist in a brain of much reason." He answered the puzzle of how marine shells come to be found in mountains by proposing they had been buried in the ground before the mountains were raised and that there was once ocean where there is now land. Da Vinci didn't publish his thoughts, recording them in his private notebooks in the early 16th century. It would be another 150 years before his views were independently revisited by Steno and Robert Hooke. Robert Hooke made a study of of fossils and geology. He was the first person to examine fossils under the microscope, observing similarities between the structures of petrified wood and fossil shells, and of living wood and living mollusk shells. He thought that the organic material could be turned to stone by immersion in water that was rich in dissolved minerals. Hooke continued to compare fossils with living organisms, concluding that many fossils represented organisms that no longer existed on Earth. This was a controversial notion at the time because the idea of extinction ran counter to religious belief.

STENO'S DISCOVERIES

In 1666, Danish anatomist Niels Stensen, known as Nicolas Steno, was given a shark to dissect. As he did so, he was taken by the resemblance of the shark's teeth to triangular pieces of rock called tongue stones. Steno declared that the tongue stones were indeed the teeth of once-living sharks and argued that the original shark tissue had been gradually replaced over time by minerals. Steno proposed that fossils were snapshots of life from different periods in Earth's history.

Steno's work on fossil shark teeth pioneered the field of paleontology.

19 Animalcules

WHEN DUTCH CLOTH MERCHANT ANTONIE VAN LEEUWENHOEK DECIDED HE WANTED TO SEE HIS FINE THREADWORK MORE CLEARLY, he ended up launching a whole new field of biology. In the process he discovered that there is a lot more to the natural world than meets the eye.

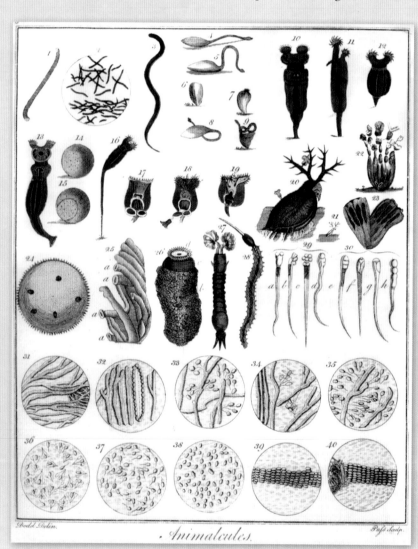

Some examples of the types of "animalcule" first observed by Leeuwenhoek.

In the 1650s, Leeuwenhoek's inventiveness led him to a way of creating small, high-quality lenses that could magnify 250 times or more. The microscopes he constructed were not barrel-shaped—their single lens was mounted between two small metal plates held close to the eye, with an adjustable pin behind to hold specimens. Using these instruments he started to examine a range of matter such as pond-water, and was amazed to discover an unsuspected world of tiny animals (he called them animalcules) swimming about. Though not a professional scientist, he made contact with the Royal Society in London and wrote the first of over 150 letters to the Society describing and illustrating his discoveries. Leeuwenhoek was practically the only person doing such research at this time, because no-one else was able to make such good lenses, and he kept his methods secret.

Antonie van Leeuwenhoek (1632–1723) is regarded as the founder of the field of microbiology. However in later life he found another calling as he became the chief wine inspector of the city of Delft.

New worlds

It would take hundreds of years for scientists to understand fully what Leeuwenhoek had drawn to their attention. He observed what we would now call protozoans (single-celled "animals"), small multicellular creatures such as rotifers, tiny plants (algae), and even bacteria. He also investigated body fluids, and was the first to see spermatozoa swimming in male semen. This discovery caused controversy—for years many scientists believed spermatozoa were parasites, rather than the sex cells we now know them to be.

20 Metamorphosis

THE TRANSFORMATION OR "METAMORPHOSIS" OF A CATERPILLAR INTO A BUTTERFLY, or a tadpole into a frog, has always excited curiosity. From the late 1600s, people started studying these remarkable changes scientifically.

René de Réaumur's illustration showing larval development in the botfly, a parasite of deer.

Dutch entomologist Jan Swammerdam made important early studies of the subject. At the time, many people thought that metamorphosis literally involved changing from one kind of animal to a completely different kind. Swammerdam disproved this, showing that developing wings were already detectable inside the larvae of insects, such as the silkworm. Swammerdam's work was greatly extended by French scientist René de Réaumur, whose *History of Insects* was published in 1734.

How does it happen?

These early studies were mainly anatomical, but attention later turned to the detailed mechanisms of the changes observed. It was noticed that organs of an insect larva are broken down at metamorphosis, and the adult's body

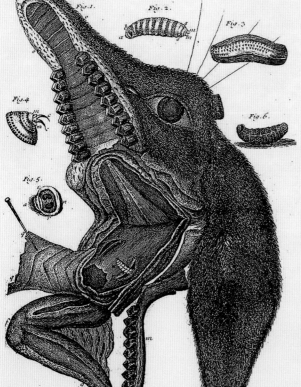

regrown from small groups of cells. After the discovery of hormones in the early 20th century, metamorphosis was shown to be under hormonal control. With frogs and toads, thyroid hormone is what triggers the change from tadpole to adult. Insects, on the other hand, have a hormone called juvenile hormone circulating in the body of the larva; when the supply of juvenile hormone is switched off, metamorphosis begins.

And why?

Metamorphosis gives an advantage to the animals concerned. It creates a division of labor—a caterpillar eats, while a butterfly devotes its energy to reproducing. It also stops competition for food between adults and their young. A tadpole feeds in water, while the frog looks for food on land. With marine creatures such as mollusks, corals, and starfish, the adults are often slow-moving or stuck in one place. These animals release thousands of tiny larvae which drift in the plankton and help the species disperse.

SILKWORM

Despite its name, a silkworm is not a worm but a caterpillar of a moth. A silkworm's life cycle shows a pattern of "complete metamorphosis," with four stages: Egg, larva, pupa (the resting stage where the adult tissues develop), and adult (or imago). As it prepares to pupate, the larva spins itself a case from fine strands of silk. These filaments are harvested and turned into fine fabric.

21 Plant Tissue

FOR A LONG TIME, KNOWLEDGE OF PLANT ANATOMY LAGGED BEHIND THAT OF ANIMAL ANATOMY. Many people thought there was nothing much worth looking at inside a plant. Nehemiah Grew set out to prove them wrong.

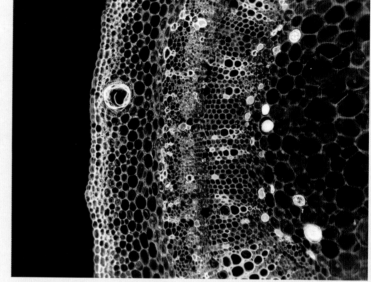

A modern microscopic image of a plant in cross-section, stained to reveal the individual tissues growing in concentric rings.

In the 1600s, microscopes were changing biology, and Grew used them to the full in his meticulous investigations of plant tissue. He published several books and articles which he collected together as *The Anatomy of Plants* (1682). This beautifully illustrated book showed for the first time how intricate the internal structure of plants is. Here was a new world just waiting to be explored. Grew wanted to make his results as clear as possible for readers, so he used what we would now call 3D cutaway drawings to display how the various tissues of plants fit together.

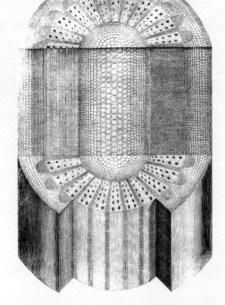

An illustration from Grew's Anatomy of Plants *showing a cutaway view of a vine branch.*

From structure to function

In this and other ways Grew was way ahead of his time, although developments in other sciences would be needed before the function of the structures he had described could be fully worked out. Although the word "cell" had been introduced by Grew's contemporary Robert Hooke to describe plant structures he saw under a microscope, Hooke was referring to empty cork cells, and there was no understanding of the cell as being the fundamental building block of living things for another 150 years. Grew himself thought of a tissue more as a structure of interwoven fibers, like a textile fabric. He appreciated the supporting function of woody reinforced tissues (which he likened to animals' bones), and proved that there were continuous tubes stretching down plant stems (we now call these xylem vessels). But it awaited work by others in the 19th century and after to fully understand the functions of plant tissues—in particular the two distinct tissues that distribute material in a plant: Xylem (which conveys water and minerals from the roots) and phloem, which transports other substances, including sugars from the leaves after they are made by photosynthesis.

In the 1720s, the English clergyman Stephen Hales investigated the function of the plant structures found by Grew. In this image from his book Vegetable Staticks, *he is investigating how liquid moves through the outer region of a woody stem.*

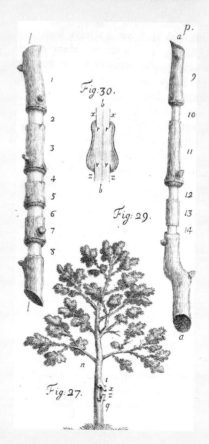

22 Flower Anatomy

WHAT IS THE PURPOSE OF FLOWERS—OR WERE THEY JUST CREATED FOR OUR PLEASURE? From the late 17th century there was renewed interest in investigating flowers more scientifically.

It had always been clear that flowers were concerned with reproduction, since they produce seeds and fruits, but exactly how does their anatomy relate to this? Early plant anatomists like Nehemiah Grew speculated about plant sexuality, and suggested that plant pollen was the equivalent of male semen in animals. But it was Grew's German contemporary Rudolph Camerarius who actually did experiments to prove this, publishing his results in *On the Sex of Plants* in 1694. For example, using plant species where each flower is either male or female (generally they are both at the same time, or hermaphrodite), Camerarius showed that female flowers would not produce seeds unless they were near male pollen-producing flowers.

The basic anatomy of a hermaphrodite flower (one that is both male and female) is shown above. Later researches clarified more details. Pollen is produced by the anthers, which are at the top of tall structures called stamens. This pollen is transferred to a sticky stigma, where it makes its way to the ovule, the female sex cell located in the base. Colorful petals have the purpose of attracting insects and other animals, which then transfer pollen from one flower to another. Specialized nectar-producing glands also attract insects. By contrast, wind-pollinated plants, such as many trees, don't need showy petals, but they produce masses of pollen, which is blown far and wide in the wind, and at least some of it blows successfully to the flowers of neighboring trees.

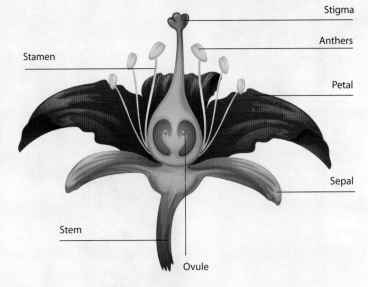

A typical flower is an organ for sending and receiving pollen, the male sex cell of a plant. Once complete, the flower will develop into a seed and fruit.

Stigma
Anthers
Stamen
Petal
Sepal
Stem
Ovule

POLLINATION

Many flowers attract bees to help them transfer pollen from one blossom to another. Some of the pollen carried by the bee brushes onto the sticky tip (stigma) of the female part of the plant. The pollen grain falls downward inside the tall female part and fertilizes the ovule or egg at the bottom. The bee harvests pollen and nectar and takes it to the hive to turn into honey.

Diagrams of floral parts, from a book about the flowers of New Zealand, published in 1888.

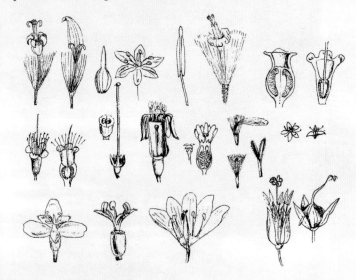

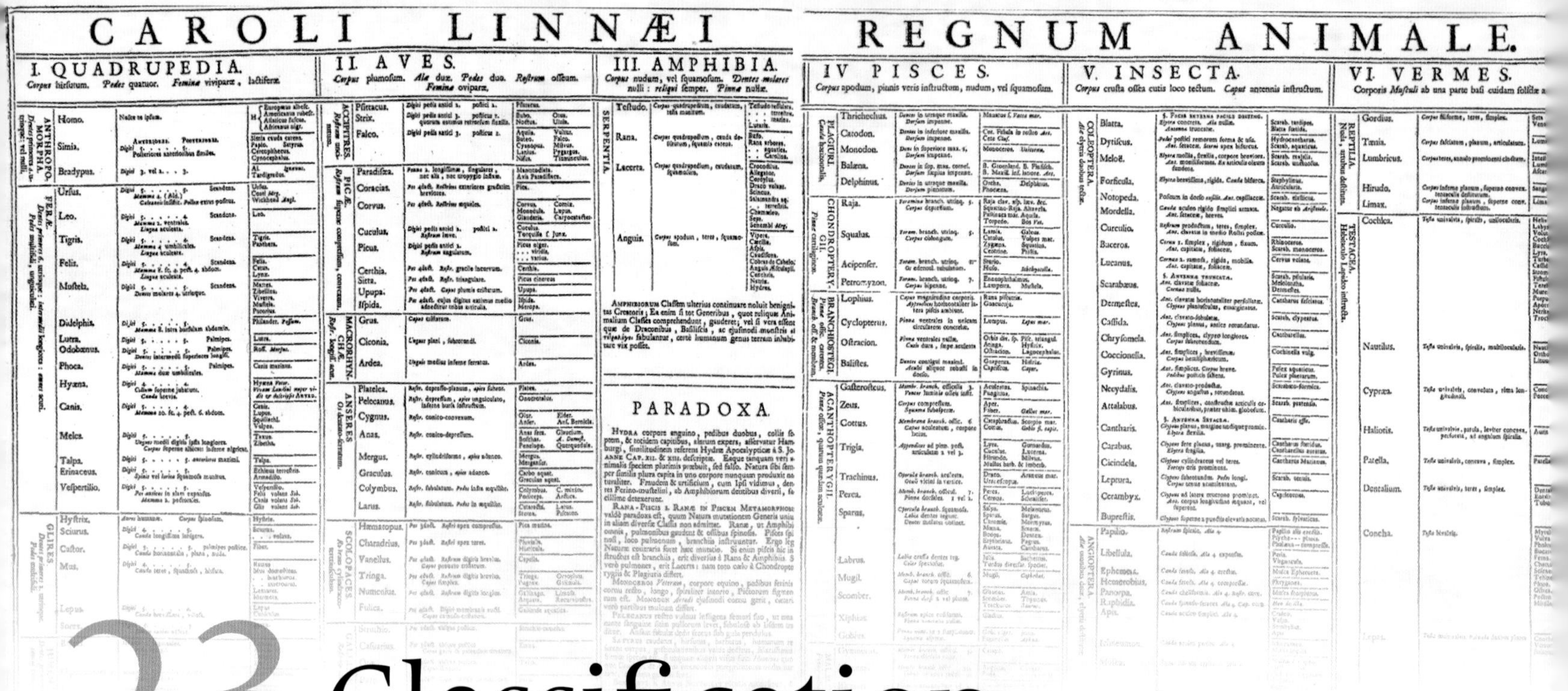

23 Classification

BIOLOGY WOULD BE CHAOS IF NO ONE KNEW FOR SURE THAT THEY WERE TALKING about the same animal or plant. Swedish naturalist Carl Linnaeus invented a method of categorizing living things that's still in use today.

Linnaeus's overall classification of animals, from the first edition of his Systema Naturae.

Linnaeus grew up with a strong interest in botany. At university he studied medicine, which then relied on growing many different herbs to treat people with. Some were known by common names—not always precise—while botany books relied on unwieldy Latin phrases for more detailed identification. Linnaeus realized that a system of short, unambiguous names was needed. He proposed that related species (or types) be given the same Latin overall name or "genus" (the Latin word for "kind"), with an adjective added for each individual species. This is called the binomial or two-name system. For example, the genus of big cats is *Panthera*. The lion is *Panthera leo* and the tiger is *Panthera tigris*. For each newly discovered species, as well as a new name, a full description would be written and a specimen kept as proof of identity.

Although he began working on it in the 1730s, Linnaeus's system took years to perfect, since he was busy as a teacher and field naturalist as well as an author. Not until 1753 for plants, in his *Species Plantarum*, and 1758 for animals, in the 10th edition of his *Systema Naturae*, did he present his new system in full.

The bigger picture

Linnaeus's binomial naming system was so effective that it quickly became popular. He was also interested in classifying animals and

BINOMIAL SYSTEM

Although Linnaeus was the first to present a consistent two-word naming system for all animals and plants, there is an earlier precedent for this approach. Swiss botanist Gaspard Bauhin (1560–1624), pictured on the right, described many plants using concise Latin phrases, sometimes only two words long, and Linnaeus re-used these on occasion. Bauhin's phrases are arguably still descriptions though, not actual names.

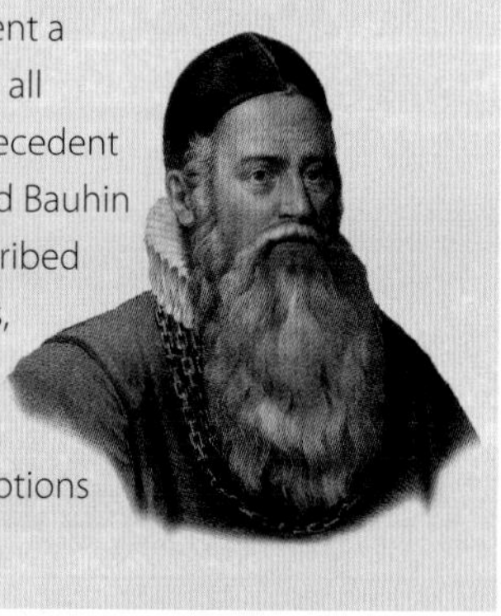

Illustrations of different bird orders from Systema Naturae.

plants on a larger scale, both to help arrange them in catalogs and to reflect their natural relationships. These two objectives don't always coincide. In the case of plants, Linnaeus chose to arrange them in an artificial manner by counting the numbers of sexual organs (stamens and pistils) in their flowers. This sometimes meant that plants that were not at all related were classified together. Linnaeus knew that it should eventually be possible to classify them more logically, as we do today, speaking about the rose family, the daisy family, and so on, but he was not attempting to create a family history of life. However, one great advantage of the Linnaean naming system is that it still works in an evolutionary context. It can also be expanded indefinitely as more species are discovered, though this has often meant splitting up his original genera. Overall, his binomial system looks set to withstand the test of time.

24 Mycology

APART FROM NEEDING TO TELL THE DIFFERENCE BETWEEN EDIBLE AND POISONOUS mushrooms, mycology—the scientific study of fungi—was little developed till recent centuries.

The researches of Italian botanist Pier Antonio Micheli (1679–1737) were a landmark in understanding these mysterious lifeforms. In Micheli's time there were lots of superstitions about fungi, including that they were created spontaneously out of rotting flesh and plant matter. Micheli focused on this question, and in his book *Nova Plantarum Genera* he not only described hundreds of species, but reported his experiments proving that fungi reproduce by tiny spores from which the new generation grows. (He showed this by growing the spores on slices of melon!) But it was still a long time before fungi were understood fully. For convenience, they were initially classified with plants, although they are quite different. They don't photosynthesize, but absorb nutrients from their surroundings, which may be decaying plant matter or a living body. A fungus's main body is a network of threads called hyphae, usually buried out of sight. Today fungi are classified in a kingdom of their own. They are actually more closely related to animals than plants!

A mushroom is the fungus's fruiting body, made from tightly packed hyphae. This one is from the poisonous fly agaric fungus. Its spores are released by the thousands from gills underneath the mushrooms' caps.

25 Botanical Gardens

DURING THE MIDDLE AGES "PHYSIC GARDENS" were created as a source of cultivated medicinal herbs. Out of these emerged the idea of botanical gardens, where plants of all kinds are studied for their own sakes.

In the 18th century, the study of botany expanded greatly, partly motivated by the desire to classify and find uses for the many exotic species being discovered in European nations' growing empires. Physic gardens, especially, such as the Jardin du Roi in Paris, France and the Royal Botanic Garden in Edinburgh, Scotland, began to expand their activities to carry out general botanical research.

The garden that was to become the world's biggest originated from the merger of two adjacent royal gardens at Kew, on the outskirts of London, England, in 1772. Leading botanist Sir Joseph Banks, who had collected plants on Captain Cook's expeditions to the South Seas, supervised its development. Kew Gardens officially became a public botanic garden in 1840. Magnificent glasshouses were built and scientific facilities extended, including what grew to be the world's largest herbarium (collection of dried plants).

GREENHOUSE

These large greenhouses at the Eden Project in Cornwall, England, are constructed as interlocking geodesic domes, so no internal supports are needed. Instead of glass they are covered with special plastic units, which are light, tough, and virtually self-cleaning. Greenhouses protect tender plants from wind and weather, and also maintain a high air temperature because their transparent coverings allow the energy from sunlight in, but stop the heat from escaping.

What botanical gardens do

Modern botanical gardens have many roles in addition to being beautiful places to visit. At the simplest level, their plants are labeled, so any visitor can find out what they are

A partial view of the huge Temperate House at Kew Gardens, which opened in 1863. It is the world's largest surviving Victorian glasshouse.

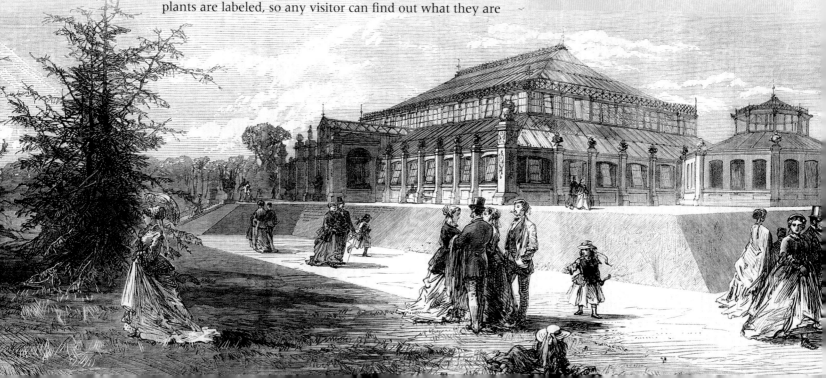

looking at. World-class gardens, such as Kew and the New York Botanical Garden, carry out a variety of teaching, conservation, and scientific activities. Many new species of plants are still being discovered, especially in the tropics, and need to be classified. Garden experts often work at compiling floras (complete inventories of plants of a particular region). "Seed banks"—preserving plants as seeds as a conservation measure—have also become increasingly important. The largest bank is located on the Arctic island of Svalbard, where the cold preserves the seeds.

KING'S GARDEN

A leading French naturalist of the 18th century, Georges-Louis Leclerc, Comte de Buffon, was head of the Jardin du Roi (King's Garden) in Paris from 1739 until his death in 1788. Although Buffon himself was more of a zoologist, he attracted talented botanists to the institution, changing the former physic and ornamental gardens into a true botanical garden that played a leading role in plant classification. After the French Revolution, the garden was renamed the Jardin des Plantes.

26 Selective Breeding

THE 18TH CENTURY SAW GREAT EFFORTS TO ADVANCE AGRICULTURE. Its achievements included the successful creation of new, improved breeds of animals and plants.

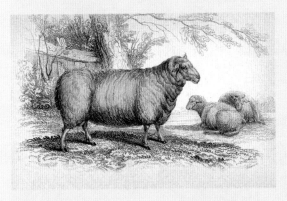

Humans have always influenced the characteristics of the animals they domesticate. Originally this would have been unconscious—tamer animals would have been easier to raise and breed, for example. But even in ancient times there were deliberate breeding programs to improve the qualities of animals such as donkeys and hunting dogs. The 1700s in Europe was an "Age of Reason," and belief in rationality extended to agriculture. Concerted efforts were made to improve existing farm animals. The most famous "improver" of the time was English agriculturalist Robert Bakewell.

Robert Bakewell (right) created new animal breeds including Leicester sheep and longhorn cattle.

By selecting which individual animals to breed from, and by crossing from other breeds, Bakewell created new breeds of cattle, sheep, and horses. He also hired out his prize bulls and rams to other farmers, very profitably. Bakewell kept his breeding methods a secret, but he probably used a combination of inbreeding (breeding between near relatives) and outbreeding. In modern terms, inbreeding is a good way to fix "good" genes in a population, so that all offspring will carry them. On the other hand, inbreeding can also cause genetic weaknesses. This was understood in a general way by Bakewell and his contemporaries, even though it was not fully described until the 20th century. Selective breeding of animals and plants remains crucial in feeding the world, and there are constant efforts to develop new varieties with greater yields and better disease resistance.

27 Respiration

WHAT IS BREATHING FOR? IT WAS ONLY IN THE LATE 1700S that scientists began to find answers to this basic question. Their work eventually led to a detailed understanding of how living things function.

"Respiration" was originally just another word for breathing. We obviously die if we can't breathe, but why? From ancient times people speculated about the purpose of breathing, though they didn't make much headway. One popular suggestion was that it cooled down the body. Further progress had to wait till another equally basic question got an answer: What is air? By the 17th century scientists knew that air had weight, but they thought it was a single substance rather than the mixture of chemical elements we recognize today. The role of air in chemical reactions wasn't understood at all.

Eminently breathable air

Eventually scientists began to realize that there are such things as different "airs," with different properties. In the early 1770s, the gas we now call oxygen was discovered independently by Englishman Joseph Priestley and Swede Carl Wilhelm Scheele. Priestley found it supported life better than ordinary air, but neither scientist realized they'd discovered a new chemical element. Priestley referred to it as "dephlogisticated air" or "eminently breathable air." Experiments showed that it makes up about one fifth of ordinary air. It was left to Frenchman Antoine Lavoisier in the 1770s and 1780s to uncover its full significance.

FATHER OF CHEMISTRY

French scientist Antoine Lavoisier was interested in chemistry from an early age. He had a genius for sorting out muddled thinking, and eventually changed the whole way of looking at chemistry, also devising a more logical nomenclature to go with his new insights. That we talk about compounds of oxygen as oxides, for example, is due to Lavoisier and his colleagues. His wife Marie-Anne worked with him, taking notes of his experiments. His great wealth from being a tax collector for the French king allowed him to construct elaborate apparatus. But following the French Revolution, he was guillotined as a traitor.

The impressive "burning glass" apparatus Lavoisier used for some of his experiments. The Sun's rays are focused by lenses to give an intense heat that could even set fire to diamonds. In this way, Lavoisier showed that pure carbon burns to make carbon dioxide, the same substance that is formed by respiration in the body.

He named the new gas oxygen, and he showed that combustion, or burning, is basically a reaction between oxygen gas and a fuel substance. Lavoisier was also the first person to compile a list of chemical elements in the modern sense, and he included oxygen in his list. In some of his experiments he compared the heat given out and the oxygen consumed by burning substances and by animals. His firm conclusion was that "respiration is a form of combustion." With his experiments and theories, Lavoisier had created a revolution in chemistry—and set one going in biology, too.

PHLOGISTON THEORY

When something burns, it's not obvious that anything is being taken from the air; it looks more like things are just being released into it. A sophisticated—but wrong—theory based on this idea dominated chemistry in the 18th century, claiming that a substance called phlogiston was given off when anything was burned. The trouble was that when you roast many metals, and they turn into what we would now call metal oxides, they weigh more than before, so phlogiston must sometimes weigh less than nothing! Lavoisier's work with oxygen junked this idea.

Later developments

Lavoisier's work raised as many questions as answers. Somewhere in the body a slow "burning" was taking place, but where, and why? Lavoisier thought it was in the lungs. Discoveries in the 19th century began to clarify the picture. It was shown that oxygen is absorbed by the lungs and transported via the arteries to all parts of the body, attached to the blood pigment hemoglobin. Deoxygenated blood carrying carbon dioxide returns to the lungs, where this gas is expelled. Much later, the chemical reactions involved were traced to within cells, and the word respiration evolved a new meaning. It now refers to all the reactions in the cell that break down food molecules to obtain energy. Usually there's a reaction with oxygen at the end of the process, but not always. Yeasts, for example, obtain energy by a process called "anaerobic respiration," meaning it burns away its fuel without the need for air.

A scene of one of Lavoisier's experiments on human respiration. On the left, Lavoisier talks to his wife Marie-Anne, who takes notes. The drawing is by Marie-Anne herself. The subject is breathing out less oxygen than he is breathing in; the missing gas is replaced by carbon dioxide. The experiment is measuring how much heat the subject produces as his respiration burns fuel.

28 Photosynthesis

PLANTS GROW AND THRIVE ALL AROUND US, BUT THEY DO NOT EAT OR MOVE OR SEEM TO DO MUCH OF ANYTHING. So how do they exist? It was a Dutch doctor living in England in 1779 who provided the first real clues.

For many centuries people didn't know where plants get the material to make their tissues from, although the most popular idea was that they got it from the soil. Then, in the 17th century, Jan Baptist van Helmont showed that even after growing a tree in a pot for five years, the weight of the soil barely changed. Van Helmont concluded that the plant made all its tissues from water. Although this is not completely wrong, only in the late 18th century did it become appreciated that plants actually get most of their substance from the air.

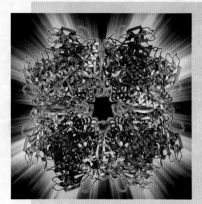

DARK REACTION

The dark reaction of photosynthesis is so called because it doesn't need light for it to take place. It is a series of chemical changes controlled by an enzyme called rubisco (short for ribulose-1,5-bisphosphate carboxylase/oxygenase). In this step, carbon dioxide from the air is combined with hydrogen and energy, delivered by the "light" reaction, to make sugars and other carbohydrates.

Light shone on the problem

Jan Ingenhousz (1730–99) was a Dutch doctor who had become well-off by performing inoculations against smallpox all over Europe. He traveled round the continent settling in different places, during which time he developed an interest in how plants work. In 1779, Ingenhousz was staying at Bowood House, a stately home in Wiltshire, England, where his friend Joseph Priestley was scientist-in-residence.

Jan Ingenhousz (right) is given some help in collecting the gas produced by his plant samples by his Austrian servant Dominique.

C4 PLANTS

Some plants have an unusual way of collecting carbon dioxide for use in their dark reaction. Known as C4 plants, they mainly grow in dry, tropical climates, like this camphor plant. Unlike most plants, they do not collect carbon dioxide during the day. That would require opening pores on the leaves to allow the gas in—but water would escape. Instead they "fix" carbon at night. It is too dark for the light reaction—but good conditions for the dark one, obviously.

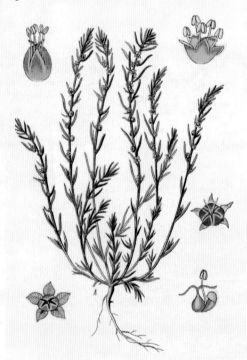

Priestley was trying to understand air, and had already shown in 1771 that—to use own his terms—air "injured" in a jar by a burning candle became pure again when a plant was afterwards grown in the jar. In modern terms, what Priestley had discovered was that plants give off oxygen, although that term was only invented a few years later by the French chemist Antoine Lavoisier.

Ingenhousz investigated this phenomenon through a careful series of experiments, published in 1779 as *Experiments on Vegetables*. He showed that plants give off oxygen when illuminated by sunlight, that it is only the green parts of plants that do this, and that in the dark plants respire like animals do, releasing carbon dioxide. Later he also suggested, correctly, that plants might take up carbon dioxide from the air to build up their tissues.

Later developments

This was the beginning of understanding the process now called photosynthesis, although that word itself, which means "making with light," was only invented in 1893. But much more work was needed to uncover what was actually going on. It was established that photosynthesizing plants make sugars for themselves out of water and carbon dioxide. The oxygen is a by-product. The plants use the sugar for energy and to make other substances. Experiments later in the 19th century showed that chloroplasts—tiny structures in plant cells that contain a green pigment called chlorophyll—are where photosynthesis takes place. The kind of light is also crucial. A plant uses chlorophyll to capture light, but it only absorbs the red and blue light, and reflects the green. That's why plants look green. (If plants could use all of the energy in sunlight, they would look black.)

Twentieth-century research has added to the story. A complex series of chemical reactions takes place, which can be simplified into two stages, the "light reaction" and the "dark reaction." In the light reaction, water molecules are split by chlorophyll using the sun's energy. The hydrogen atoms and the energy obtained from this reaction are then passed on and used in the dark reaction, which turns carbon dioxide molecules into sugar.

In an underwater plant, like this Canadian pond weed, the oxygen being produced by photosynthesis is visible as bubbles on the surface of the leaves.

29 Naturalists

INTEREST IN NATURAL HISTORY BLOSSOMED DURING THE 1700s, and the best naturalists were often enthusiastic amateurs. One of the most famous was English clergyman Gilbert White.

A view over Selborne, where Gilbert White lived. He made full use of the rich natural habitats surrounding the village.

White lived most of his life in the village where he was born, Selborne in southern England, where he studied all aspects of the local fauna and flora. Over the years he corresponded with fellow enthusiasts about his observations and conclusions, in letters later published in his classic book *The Natural History of Selborne* (1789). White believed in careful, first-hand observation rather than relying on memory or on what others had written. "The bane of our science is the comparing of one animal with another by memory," he once wrote. He described several species of mammals and birds for the first time, and researched bird migration. The work of Gilbert White and his fellow naturalists resulted in a fund of understanding that contributed to the development of disciplines including animal behavior, ecology, and animal physiology.

Gilbert White meeting a snake, from an early edition of his writings.

30 Animal Electricity

A MAJOR DISCOVERY OF 18TH-CENTURY SCIENCE WAS THE INTIMATE CONNECTION BETWEEN ELECTRICITY and the body. The early experiments in this field caught the public imagination, and eventually led to understanding how nerves and muscles work.

The role of nerves in controlling the body had been noted since ancient times from dissections and from experiments with live animals. But it was not till the 1700s that the link between nerves and electricity was made. Early experiments weren't easy to carry out, because steady currents of electricity were not available until Italian Alessandro Volta invented the electric battery in 1800. Before then, only bursts of static electricity were available—such as from a lightning conductor in a storm! Of several scientists who pioneered studies of "animal electricity," the most famous

TOPRPEDO FISH

One form of animal electricity has been well known since ancient times—that produced by torpedo fish. Using modified muscle cells, these relatives of sharks produce electric shocks to defend themselves. *Torpedo* is from a Latin word for "numb." That's what your arm feels like after you've touched one! The fish were once even recommended as a cure for headaches. Some tropical freshwater fish, including electric eels and electric catfish, also produce high-voltage shocks.

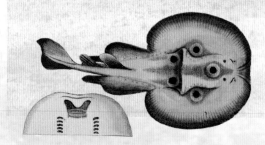

was Luigi Galvani (1737–98). His carefully described experiments, including showing how dead frogs could be made to move again, amazed the public. Some experiments seemed to show that electricity came from the animals themselves: Galvani had unknowingly turned his frogs into an electric battery by applying two different kinds of metal to them.

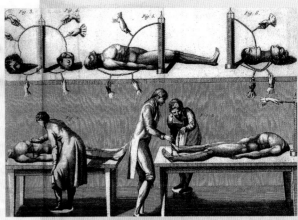

Later developments

Galvani thought his "animal electricity" might be different from other kinds, but this idea was dropped after the battery was invented, since it needed no living input. However, the phrase "galvanic electricity," meaning electricity produced by chemical means, was coined in his honor. Soon after Galvani's death, his nephew Giovanni Aldini caused a sensation at a demonstration in London by applying current from an electric battery to the dead body of a hanged murderer, which moved its limbs or assumed grotesque facial expressions, depending on where Aldini placed the electrical connections.

FRANKENSTEIN

When Mary Shelley wrote her famous novel *Frankenstein*, published in 1818, she didn't mention electricity when describing Dr. Frankenstein bringing his monster to life, though some film versions of the story have done this. However, she later acknowledged Galvani and Aldini's experiments as an inspiration for her work—especially the latter's experiments with executed criminals.

Subsequent scientists built upon Galvani and Aldini's work. German physiologist Emil Du Bois-Reymond showed in 1843 that applying a stimulus to a living nerve does indeed change its electrical state, and that this change travels down the nerve as a wave or nerve impulse. If the nerve is supplying a muscle, the impulse causes the muscle to contract. The idea that cells are the body's building blocks also led to progress. Eventually it was realized that a nerve fiber consists of many wire-like nerve cells running in parallel. In the 20th century, the biochemical cause of the electricity—the action potential—was explained.

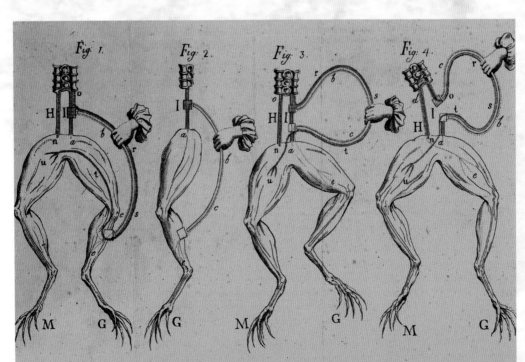

One of Galvani's experiments, showing how frogs' legs react in different ways depending on where the experimenter's metal electrodes are placed.

31 Extinction

FOR MANY PEOPLE IN EARLIER TIMES THE IDEA OF EXTINCTION SEEMED FAR-FETCHED. It was French zoologist Georges Cuvier who established more than anyone else that extinctions had really taken place.

In the early 19th century, Cuvier became an expert in vertebrate animals and their skeletons, and developed an interest in the fossils of the Paris area. Fossils were becoming accepted as evidence of past life, although what exactly they showed was disputed. Many people were skeptical about extinction. Why should God create animal species only to kill them off before humans arrived? Cuvier's own research identified fossil bones in the Paris region that clearly belonged to elephants and rhinoceroses. More than that, he showed that they were different from types alive today, and so must belong to extinct species.

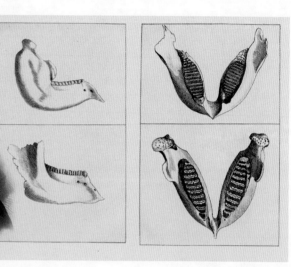

An illustration by Cuvier showing the difference between an Indian elephant's jaws and those of the extinct mammoth he had discovered.

In Cuvier's world view, animals had been created on Earth in several episodes, each ending with an upheaval or "revolution," as he characterized them, which caused their extinctions. In contrast to some of his colleagues, he did not believe in evolution, arguing that the fossil evidence did not support the gradual change of one species into another. He also believed that each species was so perfectly adapted that, if it started evolving, its parts wouldn't match each other so well.

Later, when theories of evolution became more popular, extinction was seen as part of a larger process caused by competition, climate change, and random changes. In hindsight, Cuvier was partly right: Mass extinction events, such as the one that killed off the dinosaurs, have much in common with what he proposed.

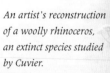

An artist's reconstruction of a woolly rhinoceros, an extinct species studied by Cuvier.

32 Biogeography

WHY ARE DIFFERENT ANIMALS AND PLANTS FOUND in different parts of the world? This apparently simple question has exercised many of the best scientific minds over the centuries.

Biogeography—the geography of living things—can be regarded as two separate though overlapping topics: Paleobiogeography and ecological biogeography. The latter explores how the distribution of animals and plants relates to present-day conditions. A pioneer in this field was German scientist and explorer Alexander von Humboldt (born in 1769). Humboldt took a holistic view of the sciences, seeking to bring together biology, climate studies, and geology to find answers to biogeographical questions. Humboldt pioneered the use of isotherms—lines of equal average temperature—on maps, which helped explain plant distributions.

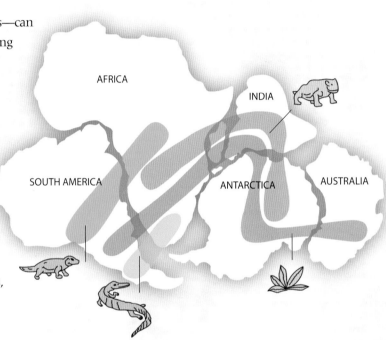

Patterns of fossils found across Gondwanaland. Gondwanaland was the southern supercontinent formed when the world continent of Pangaea broke up about 200 million years ago. Gondwanaland also later broke up, forming the southern continents we know today.

Alexander von Humboldt's daring explorations in South America—then a continent little-known to science—not only brought much new knowledge, but made him a household name when he published his researches.

Into deep time

Present-day climate doesn't explain why completely different sets of animals and plants are found in ecologically similar parts of the world, such as North Africa and Australia. This is the province of paleobiogeography, which relates the distribution of the world's living things to Earth's history. Answers to such questions are closely connected with the development of evolutionary theory. The faunas of isolated islands became a hot topic after Charles Darwin visited the Galápagos in the Pacific in the 1830s. It became obvious in hindsight that the Galápagos's strange fauna resulted largely from stray animals reaching the new islands from the South American mainland, and in time evolving into new species. Conversely, the study of fossils showed that continents now far apart had once had similar or identical fossil species. The idea that continents could move (known as continental drift) began to emerge. Humboldt was an early advocate, and later on a detailed theory was put forward by German scientist Alfred Wegener.

Wegener believed—correctly as it turns out—that all Earth's continents were once joined together in a single continent he called Pangaea, which had gradually split apart. Wegener's arguments were not fully accepted till the 1960s, when the theory of plate tectonics provided a convincing physical explanation for how continental drift could happen.

33 Lamarckism

LONG BEFORE CHARLES DARWIN, THERE WERE MANY DIFFERENT IDEAS ABOUT HOW LIFE ON EARTH MIGHT HAVE EVOLVED. The one that got the most notice was proposed by Frenchman Jean-Baptiste de Lamarck.

As the 19th century began, increasing evidence from geology and fossils pointed to the Earth being much older than suggested by the Bible, and to many strange, extinct creatures having lived before humans. Scientists began looking for explanations of how a natural process could produce the diversity of animal and plant species. Among these was the French zoologist and botanist Jean-Baptiste de Lamarck. In his book *Zoological Philosophy* (1809) he put forward his theory that changes in the body happen as a result of characteristics acquired during life. Muscles made bigger by physical work, for example, are transmitted to offspring, which would therefore start life with bigger muscles than they would have done otherwise. This idea is called Lamarckism.

As well as his theory of evolution, Jean-Baptiste de Lamarck made major contributions to studying and classifying living things, especially "invertebrate" animals, a term he coined for animals without a backbone.

Reception and debates

Some of Lamarck's colleagues, such as Georges Cuvier, thought this was a far-fetched idea, and they also thought that the fossil record lacked evidence of gradual changes between species that would be expected if evolution were true. An anonymous book was published in 1843 in Britain called *Vestiges of the Natural History of Creation*, which argued for evolution in Lamarckian terms. It became a best-seller but was criticized by scientists of the day. It was not until Darwin published *On the Origin of Species* in 1859 that a more convincing mechanism for evolution—natural selection—was put forward. This was well before genetics explained inheritance, and even Darwin thought that Lamarckian evolution might happen sometimes. A recent field known as epigenetics looks at peripheral changes to an organism's genes during life, which can be inherited. While this is not a mechanism for Lamarckism, it indicates that acquired changes may impact on natural selection.

A classic example of Lamarckism suggests that if a giraffe stretches its neck a little longer during its lifetime, that extra stretch will be passed on to its children.

34 Dinosaurs

IN 1842, MUSEUM DIRECTOR RICHARD OWEN WROTE A REPORT ON THE FOSSIL REPTILES that had been recently unearthed in Britain. He needed a new name for some of them, so made up a word from ancient Greek that meant "terrible lizard:" Dinosaur.

MARY ANNING

Mary Anning collected and sold fossils in Lyme Regis, on the south coast of England, where the cliffs are rich in fossils. She became an expert at locating and collecting extinct reptiles such as the fish-like ichthyosaur. As a woman, and without a university education, she was prevented from participating in the scientific life of her time, but members of the scientific community appreciated how crucial her discoveries and insights were.

People have come across bones and teeth of dinosaurs throughout history, but without knowing what they were. In ancient China they were called "dragon bones." With increased fossil discoveries in Europe, it was realized that huge, extinct animals, including giant reptiles, had once populated the Earth. Among the first to be unearthed were sea-living creatures including mosasaurs and ichthyosaurs. These were reptiles but in separate groups to what we call dinosaurs. The first dinosaur to be identified was discovered by English palaeontologist Gideon Mantell at a fossil-rich quarry in Sussex, England, first as a single tooth and then as a complete skeleton. This was a giant land animal with teeth like a present-day iguana, so Mantell called it Iguanodon ("iguana-tooth").

Gideon Mantell fossil-hunting with assistants at Tilgate Quarry, where the first dinosaur was unearthed.

Dead or alive?

Since Mantell and Owen's time, dinosaurs have never lost their fascination. Despite their name they are not actually lizards, but are more closely related to crocodiles. Over 1,000 fossil species are now known, and probably more are waiting to be discovered. Dinosaurs dominated the Earth from around 200 to 66 million years ago, when most became extinct, probably as a result of the giant asteroid now known to have collided with Earth at that time. Modern discoveries, especially in China, have revealed that many dinosaurs had feathers. In fact, far from being extinct, there are around 10,000 dinosaur species flourishing today. However, we call them birds.

A reconstruction of Iguanodon, a plant-eating dinosaur that could walk on either two or four legs, and had large thumb claws, possibly for defense and feeding.

35 Zoological Gardens

PEOPLE HAVE KEPT WILD AND EXOTIC ANIMALS IN CAPTIVITY SINCE ANCIENT TIMES.
Originally these "menageries" were status symbols for the wealthy, but some eventually turned into—or were replaced by—modern zoos.

The public have been enjoying exotic wildlife at the Imperial Zoo in Vienna's Tiergarten since 1752.

Menageries used to be common features of royal households. There was one at the Tower of London in England for hundreds of years, for example. With the growth of interest in natural history in the 18th century, people began to take a more scientific interest in animals. The world's oldest zoo is in Vienna. This was founded as a menagerie by the Emperor of Austria in 1752 and continues today as a modern zoo on the same site. However, the first zoological gardens planned from the start for scientific purposes were what became London Zoo. The Zoological Society of London was formed in 1826, and its Zoological Gardens opened two years later. At first the gardens were open only to members, but after 1847 the public could pay to be admitted, which helped the zoo's funding. It was in the 1840s, too, that the abbreviation "zoo" first appeared, which soon became the common name for these institutions.

The scientific use of the animals did not end when they died. On the contrary, they became available as specimens for studies in comparative anatomy, which blossomed during Victorian times.

A zoo's primary purpose has always been a place where people can see living examples of animals from all over the world. However, today they also act as a secure repository for endangered animals.

STAMFORD RAFFLES

Sir Stamford Raffles was a British colonial administrator with a strong interest in natural history, particularly of the regions in Southeast Asia where he served. On his return to London in 1824 he became a driving force behind establishing the Zoological Society of London and served as its first secretary, although he died after a few months. He is also remembered for founding the city of Singapore, on a then almost uninhabited island off the coast of Malaya.

New directions

By modern standards, conditions in traditional zoos were unsatisfactory. One advance came in 1907 at a zoo in Hamburg, Germany, which featured larger enclosures using moats instead of fences and bars. From the 1960s onwards, increasing awareness of the need for wildlife conservation, together with concerns about animal welfare, led to a change of emphasis in zoos. Today most zoos emphasize conservation and public education. They cooperate with each other in running captive breeding programs for endangered species, where possible returning some of these to the wild.

36 Vital Force

UNTIL THE 1820S, PEOPLE BELIEVED THAT A MYSTERIOUS "VITAL FORCE" underwrote life, and that the substances bodies produced couldn't be explained just by the laws of chemistry. Then an accidental discovery changed all that.

One day in 1828 German chemist Friedrich Wöhler wrote excitedly to a senior colleague that he had just made urea, a chemical known before only from urine, "without requiring a kidney." It was the first time such a bodily substance had been made in the laboratory, and heralded many more discoveries in the field of "organic chemistry" (the chemistry of carbon compounds). However, there is one way in which the chemistry of living things does differ from that of inanimate objects: Many of their molecules show "handedness"—like a factory producing only left-handed gloves. Chemists can only create "handed" molecules in a lab with some help from living things.

ORGANIC CHEMISTRY

The oil industry exploits life's ability to make complex chemicals. The first oil well was sunk in Titusville, Pennsylvania, in 1859, and today there are 65,000 oil fields being tapped around the world. Oil originates from microscopic life forms living in the distant past, and contains a mixture of organic (carbon-containing) chemicals, especially hydrocarbons (compounds of carbon and hydrogen only). As well as being used for fuel, oil provides the raw material for making a huge range of other products, from plastics to medicines.

37 Uniformitarianism

IN THE EARLY 19TH CENTURY, A NEW GEOLOGICAL THEORY AROSE THAT HELPED PERSUADE PEOPLE OF THE IMMENSE AGE OF THE EARTH. Charles Darwin accepted this new outlook, which he saw gave evolution time to happen.

Scottish geologist James Hutton was interested in geological sites showing where sediments had been laid down, hardened into rock, eroded, and then covered by later rocks. He realized the huge amount of time required for that to occur: "We find no vestige of a beginning; no prospect of an end," he wrote. Hutton's theories were extended by Charles Lyell in his *Principles of Geology* (1830), a book that influenced Charles Darwin. Lyell explained geological history using present-day processes, saying they had been occurring in the same way for millions of years. This approach became known as uniformitarianism and is still valid today, although we can see history is punctuated by occasional catastrophic events.

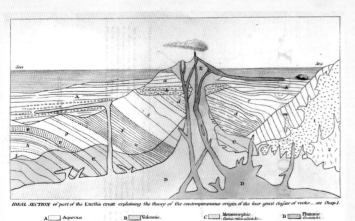

IDEAL SECTION of part of the Earth's crust explaining the theory of the contemporaneous origin of the four great classes of rocks... see Chap.1

A ☐ Aqueous. B ☐ Volcanic. C ☐ Metamorphic (Gneiss, mica schist, &c.) D ☐ Plutonic (Granite, &c.)

All the rocks older than A.B.C.D. are left uncoloured.

The frontispiece of Lyell's Principles of Geology, *showing geological processes at work.*

38 Enzymes

ENZYMES ARE BIOLOGICAL CATALYSTS THAT ENABLE AND ACCELERATE chemical reactions. They are vital to sustain life in animals and plants.

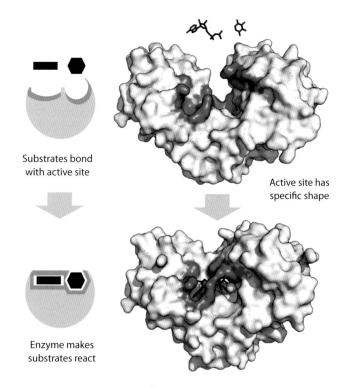

Substrates bond with active site

Active site has specific shape

Enzyme makes substrates react

The enzyme hexokinase closes over two substrates, ATP and xylose, to catalyze their bonding together, one of the reactions that takes place in cellular respiration. Areas in blue are binding sites between the enzyme and the other molecules.

The French chemists Anselme Payen and Jean-François Persoz developed a chemical from malt extract in 1833 that catalyzed the conversion of starch to sugar. They named the catalyst diastase. It was the first enzyme produced in concentrated form, although the word "enzyme" wasn't used until many years later. All enzymes discovered since then are named with the "-ase" suffix. More than 5,000 biochemical reactions are known to be catalyzed by enzymes. They operate at optimum temperatures and levels of acidity. Too cold, and the reaction rates are sluggish; too hot, and the enzyme begins to change shape and become less effective. Enzymes are generally proteins that can fold into complex shapes to accommodate other molecules, known as their substrates, which they combine or split to create new molecules, or products. In this way, certain enzymes can speed up biochemical reactions thousands of times. Almost all the metabolic processes that take place inside cells require enzymes to make reactions happen fast enough to sustain life.

In addition to discovering diastase, Anselme Payen developed processes for refining sugar and synthesizing borax from soda and boric acid. He also isolated and named the fibrous carbohydrate cellulose.

Aerobic respiration and digestion

Some of the most important metabolic reactions take place in the tiny mitochondria within a cell. In a series of reactions, enzymes facilitate the conversion of glucose and oxygen into carbon dioxide, water, and energy. The energy is needed to sustain life. Another significant set of enzymes is found in the gut. There are enzymes for all kinds of food: Fats, proteins, and starches. For example, starch is broken up into small glucose molecules, which can pass through the gut wall and into the bloodstream as an energy supply for the body's cells. Amylase enzymes are the biological catalysts for this reaction. One, called ptyalin, is produced by the salivary glands, and when saliva mixes with food in the mouth, the ptyalin helps catalyze the hydrolysis (splitting of a compound by adding a water molecule) of starch into glucose. The ptyalin continues to work on the starch in the stomach until highly acidic stomach secretions denature pytalin, or alter its shape.

39 Osmosis

OSMOSIS IS THE MOVEMENT OF A LIQUID ACROSS A SEMIPERMEABLE MEMBRANE from the lower concentration of a solute toward the higher concentration of the solute. Osmosis is a key process in water moving through living things.

TRANSPIRATION

Transpiration is effectively an internal water pump in plants. It is the process by which moisture is carried through the plant from roots to small pores on the underside of leaves, where it evaporates and is released into the air as water vapor. As water is lost at the top of the plant, osmosis pulls more up from the base. The moisture is transported through the plant in a system of tubes made of rows of special xylem cells. Transpiration is faster in warmer and windier conditions.

Osmosis was discovered in 1748 by French clergyman and physicist, Jean-Antoine Nollet, who used a pig's bladder as a crude membrane in his demonstrations. Osmosis is a form of diffusion, a process in which materials in a liquid or gas mixture naturally spread out, with the result that they become evenly distributed throughout the mixture. In osmosis, the membrane blocks the motion of larger molecules, but water can move through. The water moves to even out the concentrations of whatever is in the mixture—moving from the low concentration to the higher side. The osmotic effect stops when the concentrations are even on both sides of the membrane, but that is achieved by having a larger volume of water on one side.

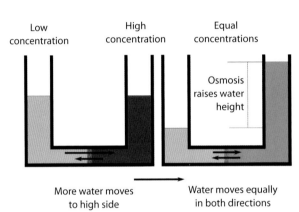

Low concentration   High concentration   Equal concentrations

Osmosis raises water height

More water moves to high side   Water moves equally in both directions

Biological membranes are semipermeable and will not let larger particles such as proteins through. Charged ions cannot pass either. Small molecules, such as oxygen, carbon dioxide, nitrogen, and nitric oxide pass through by diffusion. Osmosis is the primary means by which water is transported into and out of cells.

Cut flowers are placed in hypotonic water (fresh, clean water will do) in order to push water into their stems by osmosis. The process keeps the flowers looking perky and fresh for longer—but eventually the dead plants will succumb to the natural processes of decay. A living plant wilts when it lacks water, and can be revived with water.

In biological systems, water with a lower concentration of dissolved substances than normal is hypotonic; a high concentration is hypertonic. If a cell is in a hypotonic solution, osmosis makes water moves into the cell and the cell becomes swollen. This will happen if the cell is in fresh water, for example. If the cell is in a hypertonic solution, such as saltwater, water molecules will move out of the cell, which will become shriveled. Plants use osmosis to draw in water from the soil. Solutes are concentrated in root cells, so water molecules pass into the cells, increasing the pressure within and making them firm. If a plant is deprived of water for a long time, the pressure inside cells falls; the cells shrivel and the plant wilts.

40 Cell Theory

CELL THEORY IS THE FOUNDATION OF MODERN BIOLOGY. It is guided by the principle that cells are the fundamental units of all living organisms and can only arise from pre-existing cells.

The development of the microscope in the 17th century laid the foundations for cell biology as a scientific discipline. For the next century, scientists continued to observe animal and plant cells under increasingly more powerful microscopes. Biologists had long believed that living organisms were made up of fundamental units, but no one made the link between the cells they were seeing under microscopes, and the cells that made up the bodies of animals and plants. French physiologist Henri Dutrochet finally made the connection in 1824, when he recognized the significance of individual cells in the function of a living organism, stating that "the cell is the fundamental element of organization."

Matthias Schleiden was a professor of botany at the University of Jena in Germany.

TENETS OF CELL THEORY
1. **All living organisms are composed of one or more cells.**
2. **The cell is the basic unit of structure and organization in organisms.**
3. **Cells arise from pre-existing cells.**

Cell theory

Two scientists are credited with formulating the principles of cell theory. In 1838, German botanist Matthias Schleiden published *Contributions to Our Knowledge of Phytogenesis*, based on his microscopic studies of plants. Schleiden proposed that all parts of plants consisted of cells. He also noted the importance of the nucleus in cell division, recognizing that new plant cells formed from the nuclei of old plant cells. Also in the 1830s, the German physiologist Theodor Schwann was working exclusively with animal tissue, observing animal cells under the microscope, especially nerve cells, and noting their properties. When the two scientists met to discuss their work, and Schleiden explained the role of nuclei in cell division, Schwann immediately recognized similar structures in animal cells and realized the link between the two. In 1839, Schwann published *Microscopical Researches*, in which he stated "All living things are composed of cells and cell products." The two scientists thus became the first proponents of cell theory, resulting in two of the three basic tenets of the theory: 1. All living organisms are composed of one or

This illustration shows the structure of a typical plant cell. Unlike animal cells, plant cells also include a cell wall, chloroplasts, and vacuoles.

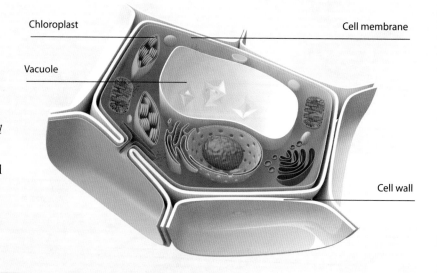

Chloroplast

Cell membrane

Vacuole

Cell wall

The illustration on the right shows the structure of a typical animal cell. Inside the membrane is a watery liquid called cytoplasm. The nucleus at the center of the cell consists of genetic material in the form of DNA. Other structures include organelles, such as the mitochondria, which produce energy for the cell, and the endoplasmic reticulum and Golgi apparatus, which produce the materials the cell needs.

more cells. 2. The cell is the basic unit of structure and organization in organisms.

Building on the work of Schleiden and Schwann, German scientist Rudolf Virchow added a third tenet: 3. Cells arise from pre-existing cells. While Virchow is credited for this contribution, he based his conclusion on the experiments of German biologist Robert Remak, who observed cell division in fertilized frog eggs.

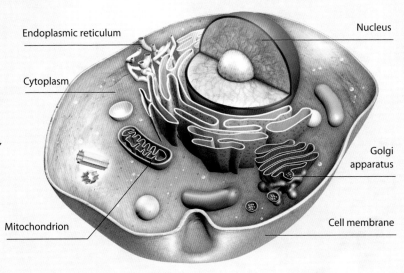

Endoplasmic reticulum

Nucleus

Cytoplasm

Golgi apparatus

Mitochondrion

Cell membrane

Properties of cells

Animal cells and plant cells have many common properties. Both have nuclei, which contain the genetic material that controls all the activities inside cells. Both also have cell membranes, which regulate the movement of water and dissolved chemicals into and out of the cell. The membrane surrounds the liquid interior, known as the cytoplasm. But, unlike animal cells, plant cells have some other, unique structures. These include cell walls made from cellulose fibers. This strengthens the body of the cell and allows cells to be cemented together like bricks in a wall. Plant cells also have vacuoles (storage organelles), which are filled with water to keep the cell firm. The cells in the green parts of plants contain chloroplasts. Within these structures is the green pigment chlorophyll, which absorbs the energy of sunlight to make food in a process called photosynthesis.

Virchow's cell illustrations made for a series of lectures given in 1858.

Theodor Schwann developed the basic principle of cell theory— that all living things are composed of cells. His contribution to the classification of different cells laid the foundations for modern histology.

Function of cells

While all cells share characteristics, a body contains many specialist cell types. A good example is the red blood cell. This is red due to the presence of hemoglobin, a molecule that carries oxygen around the body. Nerve cells have a very different, wire-like structure (for sending electrical signals), while bone cells are surrounded by a network of calcium phosphate crystals, which gives the bone its rigidity. All these cell types arise from unspecialized stem cells, which can transform, or differentiate, into any cell type needed in the body.

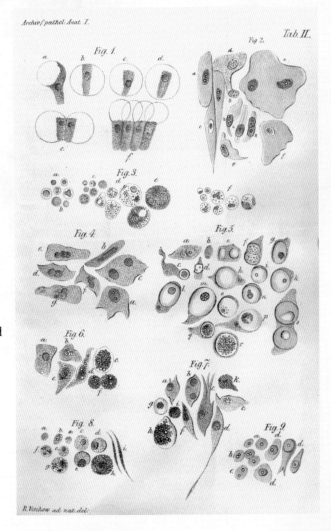

41 The Heart

THIS MUSCULAR ORGAN PUMPS BLOOD AROUND THE BODY. IT IS POWERED BY THE CARDIAC MUSCLE, which keeps it beating from conception to death, typically about 100,000 times every day.

The human heart has four chambers: Two atria and two ventricles. The left side of the heart receives oxygen-rich blood from the lungs and pumps it to all parts of the body, where the oxygen is used in cellular respiration. The right side receives oxygen-depleted blood that has drained from all parts of the body, and pumps it to the lungs to pick up more oxygen. The two sides of the heart are separated by an inner wall called the septum. While the English anatomist William Harvey had studied the structure of the heart and the circulation of blood around the body, it was the Czech physiologist Jan Evangelista Purkinje who discovered the fibrous tissues (now called Purkinje fibers) that conduct electrical impulses to the heart's ventricles, stimulating their synchronized contractions to produce a complex heartbeat known as the cardiac cycle.

JAN EVANGELISTA PURKINJE

In the mid-19th century, Purkinje was one of the world's best-known scientists. Apart from demonstrating the role of Purkinje fibers in the heart, he also discovered the large neurons in the brain (Purkinje cells) and introduced the terms plasma, for the liquid component of blood, and protoplasm, for the content of a cell. Purkinje also established the world's first department of physiology in 1839, at the University of Breslau, Poland.

Two circulations

There are two circulation systems in the body: The pulmonary circulation transports blood to and from the lungs, while the systemic circulation carries blood to and from the rest of the body. The right atrium receives oxygen-poor blood via two large veins, the inferior vena cava (from the lower and middle body) and the superior vena cava (from the body above the diaphragm). From the right atrium, the blood travels through the tricuspid valve into the right ventricle. The heart's valves prevent blood from flowing backward. From the right ventricle, it is pumped up through the pulmonary valve and into the pulmonary trunk, which branches into the two pulmonary arteries, each connecting to the lungs. The arteries branch into smaller

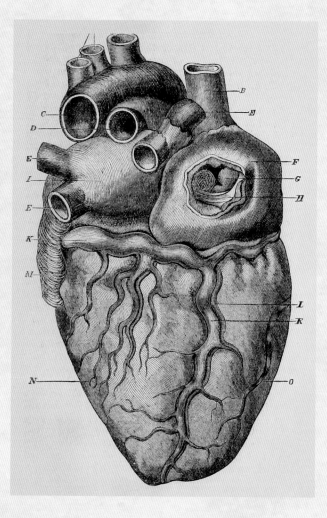

A diagram dating from 1881 of the main blood vessels and chambers of the heart.

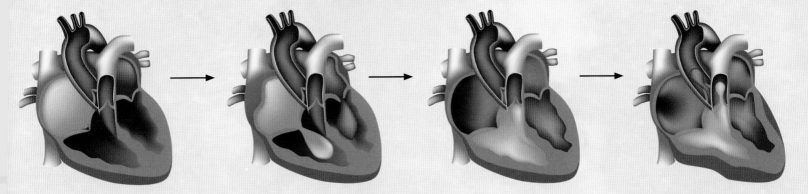

and smaller blood vessels in the lungs, where they pick up oxygen from alveoli. The contraction of the right ventricle is not strong enough to pump the oxygenated blood all around the body after the lungs, so the oxygen-rich blood has to return to the heart to be pumped again. It travels in four pulmonary veins, two from each lung, and enters the left atrium. Continuing through the mitral valve, it passes to the left ventricle. The muscle enveloping this chamber is thicker than that of the right ventricle because it has to pump the blood to the whole of the body—not just to the lungs. From the left ventricle, the oxygen-rich blood passes through the aortic valve and leaves the heart via the aorta, the largest artery in the human body.

The cardiac cycle involves a wavelike contraction of muscle with each beat of the heart. Blood arrives in both atria at the top of the heart, and the heart muscle begins to contract here, forcing blood into the ventricles below. The contraction then reaches the bottom of the heart, pushing blood up the arteries and out again.

42 Alternation of Generations

SIMPLE PLANTS HAVE A LIFE CYCLE THAT INVOLVES A GAMETOPHYTE STRUCTURE, which produces sex cells, followed by a sporophyte structure, which produces spores asexually. This is called the alternation of generations.

Mosses are very simple plants without true leaves, stems, or roots. The most visible parts are usually the gametophyte generation, which is involved in sexual reproduction. The sporophyte generation is concerned with dispersal.

The German biologist Wilhelm Hofmeister first outlined the alternation of generations in 1851, but advances in genetics were needed to fully understand it. The sporophyte is the diploid phase of the plant, meaning its cells contain a double set of genes, one from each parent. The sporophyte produces spores, tiny units that are dispersed by wind or water. The spores are haploid, which means they contain just a single set of genes, and they grow into the gametophyte structure. This produces gametes (eggs and sperm, both also haploid), which fuse to create a single diploid cell—which then becomes the sporophyte. In mosses and liverworts, the dominant generation is haploid, so that the gametophyte comprises what we think of as the main plant. In plants with internal veins, such as ferns, the sporophyte is the main plant.

43 Darwinism

IN THE WORLD OF IDEAS, 1859 WAS A YEAR UNLIKE ANY OTHER. It saw the publication of Charles Darwin's theory of evolution, which revolutionized scientific thinking and our understanding of life on Earth.

Charles Darwin was born in Shrewsbury, England, into a wealthy family. The son of a doctor, he studied medicine for two years but then gave it up. Charles then studied at Cambridge University with a view to becoming a clergyman, and, like many gentlemen scholars of his age, he developed a keen interest in botany, entomology (insects), and geology.

In 1871, Darwin's book The Descent of Man *caused controversy by suggesting that humans and apes, such as chimpanzees, were descended from a common ancestor. Critics misinterpreted this, claiming Darwin thought people were descended from modern monkeys.*

Voyage of the *Beagle*

On graduating from Cambridge, Darwin heard that a ship called the HMS *Beagle* was about to embark on a scientific expedition around the world. Darwin applied for the post of unpaid naturalist (in fact he even had to pay his own way) and was accepted. Sailing in December 1831, the young man suffered seasickness on the Atlantic crossing, but delighted in the tropical rainforest when the *Beagle* docked on the coast of Brazil.

The Beagle *laid ashore for maintenance on the Santa Cruz River in Argentina. Further down the coast, Darwin marveled at fossils of a giant sloth, and pondered its extinction.*

Rounding South America's southern tip, the *Beagle* sailed north, reaching the remote Galápagos Islands off Ecuador in 1835. These small islands were home to extraordinary species found nowhere else, such as marine iguanas, flightless cormorants, giant tortoises, and even penguins that lived far from the Antarctic. Darwin saw that different species of animals, for example tortoises, existed on various islands, differing by features such as the shape of the shell. The young naturalist also collected finches with beaks of different shapes from different islands. These experiences gave him much to think about as the *Beagle* sailed on.

This Victorian image of battling dinosaurs illustrates the idea of "the survival of the fittest." However Darwin himself did not use this phrase; it was coined by a supporter, Herbert Spencer.

Natural selection

Back in England, Darwin's letters describing his experiences had impressed the scientific community. Before long he had married his cousin, Emma, settled in Downe in Kent, and started a family. Meanwhile, he pondered what he had seen on the Galápagos and elsewhere on his voyage. Could it be that the little birds were all descended from a common ancestor that had reached the islands? Did this single species diversify in response to varying

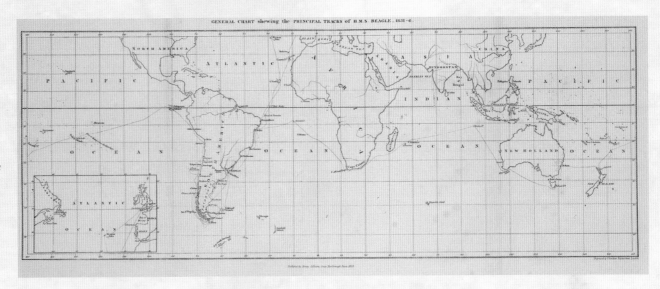

The Beagle's five-year voyage (1831–36) explored the coast of South America, then crossed the Pacific, visiting Australasia before returning to England via the Indian Ocean. Darwin later wrote: "The voyage of the Beagle has been by far the most important event in my life and has determined my whole career."

conditions on different islands? Darwin wrote to scientists and animal breeders to gather evidence for his growing conviction that evolution occurred through a process he called natural selection. Every generation of animals produced more young than could survive. This meant that all were in competition. Yet all individuals within a species were not exactly alike. Natural variation meant that some were better suited to their environment and more likely to reproduce. For example, those with slightly longer legs could run faster from danger. These individuals were most likely to survive long enough to breed, and so pass on their characteristics. Over time, a favourable trait would gradually spread through the whole population, and the species would change, or evolve.

Going public

Darwin was well aware that his idea would cause uproar among scientists and 19th-century society at large. So he delayed publication, instead gathering a vast amount of data to support his theory. This situation continued for 20 years, until 1858, when he received a letter from the naturalist Alfred Russel Wallace, who was working in Southeast Asia. Wallace had come up with the same idea! The following year, Darwin rushed out the book *On the Origin of Species by Means of Natural Selection*, and as he had feared, Darwin's idea did indeed cause a furor. Debate over the theory continues.

Darwin could not explain exactly how characteristics were inherited. But the 20th century brought the new science of genetics, which explained the mechanism of heredity. Darwin's theory has now become one of the foundations of modern biology.

SEXUAL SELECTION

Darwin developed the theory of sexual selection, whereby animals select their mates based on certain characteristics that do not simply favor individual survival. For example, male peacocks attract females with their spectacular, but cumbersome, tails. Since peahens choose the males with the most elaborate plumage (bright colors show the males are healthy, fitting mates), the tails enable the peacocks to reproduce.

44 Mendel's Laws of Inheritance

IN 1865, THE WORK OF THE AUSTRIAN MONK GREGOR MENDEL laid the foundations for modern genetics. Mendel's painstaking studies of pea plants explained the principles of heredity.

Mendel studied seven different traits of pea plants to come up with his laws of inheritance. They were: The shape and color of the peas, the color of the flowers and their position on the plant, the shape and color of the pea pods, and the length of the plant's stems.

Farmers had been selectively breeding crops and livestock for centuries before Gregor Mendel (1822–84) conducted his experiments in the gardens of his monastery in Moravia. Farmers bred from individuals with desirable traits based on the principle that offspring inherited these traits from their parents. But no one understood what rules governed inheritance until Mendel uncovered some in 1865. In that year, Mendel published the results of seven years of experimentation with pea plants.

Mendel selected plants that differed in a single trait, such as flower color or whether the pea was wrinkled or round. He then cross-bred his selected plants with each other, counting the number of plants exhibiting those traits in future generations. He noticed that the first generation of cross-bred plants, called F1, meaning "first filial," was comprised entirely of plants with only one of the traits. In a cross between plants with red (actually nearer purple, but the point remains the same) and white flowers, all of F1 had red flowers. There was no blending. Mendel described red flowers, and other traits that appeared in F1 as "dominant," while those that disappeared, such as white flowers, were "recessive."

Mendel self-fertilized the F1 pea plants. (He used the pollen from the same flower to fertilize the seeds—peas can do this; but many plants cannot.) He found that roughly three-quarters of the next generation, F2, had

XVII, 3. 106. Leguminosae.

453. Pisum sativum L.   Brech-Erbse.

CO-DOMINANCE

Mendel was wise in his choice of pea plants because they are quick to reproduce and have distinct traits. Genetics is not always so simple. In some cases, the alleles for a gene are co-dominant or incompletely dominant. Incomplete dominance occurs when alleles result in a blend of the two different inherited traits, for example when a red and a white flower produce a pink one. In cases of co-dominance, both inherited traits appear in the offspring. The fur color of some animals, such as horses and cats, is controlled by this kind of inheritance, which results in many varied colors and patterns.

the dominant trait, and a quarter had the recessive one, making a ratio of 3:1.

Moving factors

Mendel suggested that traits were determined by units of inheritance, which he called "factors." Factors were being transmitted from parents to offspring, so every offspring had two factors. Every trait had a factor, but they existed in different versions, such as red and white for flower color. The dominant factor would control what trait the offspring showed. Mendel also realized that factors did not merge or combine in the F1 generation and were passed on separately to F2.

He summarized his observations in two laws of inheritance: The law of segregation states that for any trait, each parent's pairing of factors separate, and only one passes from each parent on to its offspring. It is a matter of chance which version in a pair gets inherited. The law of independent assortment states that different pairs of factors are passed to offspring independently of each other.

Gregor Johann Mendel lived at St. Thomas's Abbey in Brno, Moravia, now in the Czech Republic, but then part of the wider Austro-Hungarian Empire.

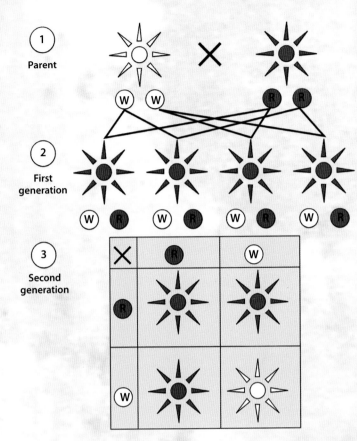

This illustration shows the passage of dominant (R) and recessive (W) alleles through three generations.

Modern view

We would now call Mendel's factor a "gene," and the different versions of a gene are its alleles. These modern terms came almost five decades after Mendel's work, which lay ignored until the start of the 20th century. Mendelian inheritance is the basis of our understanding of genetics, although in other situations, genes do not split easily into dominant and recessive forms (see Co-dominance box). However, with the benefit of modern genetics, we can explain what is going on. It is easier to picture Mendelian inheritance by giving each allele a symbol: W for white and R for red. The parent plants are WW and RR; they have two copies of the same allele for flower color. Plants in F1 inherit a W and an R, making a WR. R is always dominant, so all of F1 have red flowers. F1 plants pass on either a W or an R to F2, entirely at random. Therefore it is possible for an F2 plant to inherit RR, WR, RW, or WW. The chance of inheriting one of these four sets is equal. However, only the WW set results in white flowers; the other three create red flowers. So if there is a significantly large enough group of crosses, the ratio of red to white flowers will always be 3:1.

BAD BEES

Mendel is most famous for his experiments with pea plants, but he also bred bees to determine genetic traits in animals—as well as to produce excellent honey for monks at the monastery. It has been suggested that Mendel successfully bred a new hybrid strain of honeybees, but the bees he created were so aggressive that they stung everything in sight. Legend has it that the hive was so out of control that the bees had to be destroyed!

45 DNA

DEOXYRIBONUCLEIC ACID, OR DNA, is the most iconic chemical in the world. It was discovered back in 1869 by a Swiss doctor named Friedrich Miescher, who was studying pus!

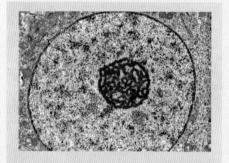

THE NUCLEUS

DNA was discovered by analyzing the contents of the cell nucleus. The nucleus of a cell is a central body bounded by two membranes. It was the first body to be seen inside cells, and it is now known that the nucleus serves as a store for DNA. The DNA contains sections of code, called genes, which are copied and taken to other parts of the cell for reading.

Biologists had long believed that there was some "genetic" entity that traveled between generations making offspring resemble their parents. Charles Darwin relied on this idea to explain his theory of evolution. However, nobody knew what it was. The first step in finding out came when Miescher investigated proteins in leucocytes (white blood cells). He harvested the cells from an unpleasant source: Pus-soaked bandages from patients with infections, gleaned from a nearby clinic. In 1869, Miescher noticed a mystery substance that did not resemble the other proteins in cell nuclei. He named it nuclein; we now call it nucleic acid. Miescher suggested these chemicals were responsible for heredity but he had no evidence for that. In the 20th century, scientists confirmed that DNA molecules did pass on from one generation to the next and were responsible for inherited characteristics. But, although it was known that DNA was made of smaller units of ribose sugars and four nucleic acids, no one knew how they fitted together. Solving that puzzle would reveal how a chemical could carry the genetic code.

Friedrich Miescher called the mystery substance he found in the nuclei of cells nuclein. The original term is reflected in the modern name of deoxyribonucleic acid, or DNA for short. Miescher went on to become Professor of Physiology at the University of Basel.

46 Conservation

CONSERVATION IS THE PROTECTION OF NATURE, NATURAL RESOURCES, AND WILDLIFE. Today conservation is carried out worldwide by thousands of organizations, but it all began with the effort to save a patch of wilderness in the Rocky Mountains.

RED LIST

The International Union for the Conservation of Nature (IUCN) publishes the Red List, a comprehensive guide to wildlife at risk. It classifies plants and animals within seven categories, from Least Concern through Endangered to Extinct.

As a popular movement, conservation started in North America in the mid-1800s. Earlier, as European settlers spread west across the continent, nature had been seen as something to be conquered, tamed, and exploited. Then a few thinkers, such as John Muir and Henry David Thoreau, realized that natural resources and wildlife were not limitless, but needed to be protected. In 1871 geologist Ferdinand Hayden led the first survey of northwest Wyoming. He recommended that a national park be established to protect the unique scenery he saw in Yellowstone, which included

geysers and hot springs. Many local people opposed the move on the grounds that it would hinder forestry, hunting, and mining. However businesses such as the Northern Pacific Railroad supported it because it would open the region to the infant industry of tourism. In 1872, a law was passed making Yellowstone the world's first national park, "a pleasuring ground for the benefit and enjoyment of the people." The word conservation, from the Latin *servare* "to guard," and *con*, "together," was probably coined by Gifford Pinchot, head of the U.S. Forest Service in the early 1900s.

In 1871 the geological survey led by Ferdinand Hayden passed Mirror Lake on its way to the East Fork (now called the Lamar River) of the Yellowstone River. Hayden's report recommended the region be protected as a national park.

Conservation in the 20th century

The late 1800s saw the creation of more national parks in the USA, and also in Canada, Australia, and Europe. In the 1950s, the national park movement gathered pace. There are now national parks and reserves on every continent, and some are the size of small countries. National parks have two main aims: To preserve nature and allow people to enjoy it. With growing numbers of visitors, parks have to be carefully managed to ensure these aims do not conflict. Modern conservation is carried out on many fronts. Conservationists work to preserve natural resources such as forests, soil, and water supplies. They also act to protect rare wildlife, for example by banning the hunting of rare species. Almost 150 years after Yellowstone was set up, preserving whole ecosystems is still seen as the best way of protecting wildlife.

Yellowstone National Park protects 8,990 square kilometers (3,470 square miles) of spectacular scenery, including mountains, canyons, waterfalls, geysers, and hot springs. The last two features are caused by volcanic activity.

47 The Biosphere

LIFE EXISTS ON EARTH ON DRY LAND AND IN THE OCEANS, IN SOIL AND ROCKS, and high in the atmosphere. The region of a planet that sustains life is called its biosphere. This term, now central to ecology, was the brainchild of an Austrian geologist named Eduard Suess.

Austrian geologist Eduard Suess proposed the term biosphere in 1875. In the 1920s, his idea was developed by Russian geochemist Vladimir Vernadsky, who defined ecology as the science of the biosphere.

On Earth, the biosphere extends from the Arctic to Antarctica, and up into the lower atmosphere, where microbes have been detected 64 km (40 miles) up in the air. Life is found throughout the oceans, including in trenches plunging to 8,530 m (28,000 ft), and single-celled organisms are found in the soil and rocks at least 19 km (12 miles) below ground. However, the ocean depths are dark and cold, and the upper atmosphere contains little oxygen. So, most life exists in a relatively narrow band from 500 m (1,640 ft) below the ocean surface to 6 km (3.75 miles) up in the atmosphere. This layer is incredibly thin compared to the size of Earth. If an apple were as big as the Earth, the biosphere would be the skin around it. To date, Earth is the only planet we know for certain sustains life in a biosphere. Energy from the Sun, water, and an atmosphere containing oxygen are needed for life as we know it, and our planet orbits the Sun at just the right distance for temperatures to be comfortable. In future, we may detect biospheres on other planets or moons in our solar system, and orbiting other stars. Will they harbor life?

The air, soil, rocks, glacier, and lake in this alpine scene all contain life, and so are part of the biosphere.

48 Germ Theory

WE ARE NOW WELL AWARE THAT INFECTIOUS DISEASES are the result of other organisms invading our bodies. However, that fact was not proven until a series of experiments in the 1860s involving some strange glassware.

The link between cleanliness and good health is an old one. The Romans had public baths in every city for the public to keep clean, and early Islamic teaching stressed the importance of regular washing. However, no one knew what it was about dirt that caused disease. Until the 19th century, Western doctors mostly put it down to miasma, or "bad air," which emanated from rotting bodies and bodily waste. Accordingly, the best way to avoid catching a disease, such as cholera or plague, was to wear an elaborate beak-shaped mask, packed with herbs and minerals that was supposed to absorb and filter out the deadly miasma. Despite this protection, doctors frequently fell ill.

Industrial influence

While there was no doubt that dirt would lead to diseases, including deadly infections of wounds following surgery, the true cause was not revealed by a doctor, but by a chemist. In the late 1850s, Frenchman Louis Pasteur was asked to find out why wine and other drinks sometimes spoiled in the barrel and bottle. Pasteur found that this was due to fermentation from yeasts and other microorganisms mixed into the liquid. Moreover, he showed that this yeast—found naturally on fruits—was essential for making wine and other alcoholic beverages. If it were removed, the grape juice would not transform into wine at all. This finding contradicted the accepted theory at the time that yeast and other microbes formed from non-living material spontaneously. Pasteur performed his most famous experiment to prove that this was not the case. He boiled up a broth to kill all microorganisms and placed it in a series of flasks. Some were sealed entirely from the air, others were connected to the air via convoluted "swan-necked" tubes, and others were left open. The sealed flasks did not spoil, while the open flasks went bad after a few days. The swan-necked flasks took a lot longer to spoil. Pasteur presented this as evidence that the cause of the spoiling was from "germs" carried in the air, not arising out of the broth itself. It had just taken the germs longer for them to get in via the curved tubes. Pasteur's "germ theory" had far-reaching effects. It showed that rotting and putrefaction were caused by microorganisms, not some kind of reverse process where rotting material created germs. In addition, the all-pervading germs were soon linked to infections in humans and animals.

JOHN SNOW

In 1854, an epidemic of cholera hit Soho, which was at that time a slum district in London, England. Cholera is a nasty digestive disease, and the Soho outbreak killed 600 people. A local doctor, John Snow, mapped the disease and found that the victims all collected water from the same pump on Broad Street. Snow found the pump's supply was mixing with a nearby cesspit, thus proving that the dirty water was carrying the cause of the disease.

Pasteur developed a method for removing microorganisms from drinks, such as wine and milk. It involved heating them to high temperatures for a fraction of a second to kill the pathogens, but leaving the products largely unaffected. This process is now called pasteurization.

A replica of Pasteur's swan-necked flask, which has now been sealed from the air, contains some broth. Would you drink it?

Pathogenic organisms

In the 1870s, the German microbiologist Robert Koch took Pasteur's ideas further by showing that a particular disease was caused by a specific germ. He grew samples of bacteria on agar, a nutrient-rich gel derived from seaweed, which allowed him to isolate and study different strains. Koch identified the different bacteria that cause anthrax, a disease of livestock that also affects humans, the lung disease tuberculosis, and cholera. Organisms that cause disease are now described as pathogens. As well as bacteria, pathogens include viruses, protists, such as amoebas, and even worms.

49 Microbiology

MICROBIOLOGY IS THE STUDY OF MICROSCOPIC, usually unicellular, organisms, including bacteria and protists.

The study of microorganisms began with the invention of the microscope. With this new invention, Antonie van Leeuwenhoek was able to observe bacteria in the late 17th century; however, the German Robert Koch is recognized as the pre-eminent "father of microbiology." In the 1870s, he began to show which bacteria or other microbes were responsible for dangerous diseases.

The Petri dish was invented by Julius Petri, who further developed the technique of agar culture to cultivate bacterial colonies derived from a single cell. This method proved to be invaluable in bacteriology.

Robert Koch at work, sitting at his microscope in the early 1900s. Koch's methodology for isolating the causes of infectious disease is still used today.

Two groups

The two most important groups studied by microbiologists are single-celled bacteria and protists. A bacterial cell is many hundreds of times smaller than the average human cell. Many bacteria live in and on the human body; there are said to be 10 bacteria for every human cell, with almost all being harmless, or even helpful, although a few will cause diseases. Bacteria are prokaryotic, which means they do not have a nucleus or other organelles in their cell. Bacterial cells generally have a rigid, protective wall around the cell membrane that surrounds the contents. They may also have tail-like flagella used for swimming, and hair-like extensions called pili on the outside, connecting to other cells or holding on to a surface. Bacteria-like prokaryotes, called archaea, look similar but have a distinct evolutionary heritage.

Protists are eukaryotic organisms. That means they have larger cells than bacteria, and their cells contain complex structures, including a nucleus. Plants, animals, and fungi are multicellular eukaryotes. But a protist's body is just one cell—or in a few cases a colony of cells—and they are highly varied organisms, such as protozoa, algae, and amoebas. Some protists have one or two flagella for propulsion, while some use cilia, hair-like extensions that waft to create currents around the cell. Others move by deforming their cell. Some protists are more animal-like, in that they consume food, while others are plant-like and can photosynthesize. Although single-celled, yeasts are characterized as fungi.

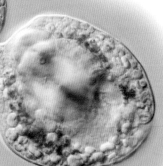

These euglenids are protists that have the best of both worlds. The orange blobs are used for photosynthesis, which supplies the cell with sugar. However, the euglenids are also active hunters that engulf and digest any food they find.

50 Nitrogen Cycle

ALL PLANTS AND ANIMALS NEED NITROGEN TO MAKE PROTEINS AND DNA.
Most organisms cannot access the pure nitrogen gas in the air and instead
rely on a nitrogen cycle that supplies life with a usable form.

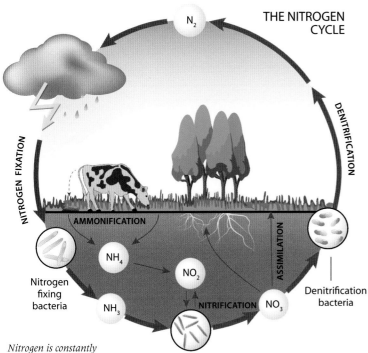

THE NITROGEN CYCLE

Nitrogen is constantly cycling through the biosphere as it is used by life to make essential compounds.

Nitrogen is crucial for all life, and the food
chain depends on it. The problem is that
although it is an abundant element, its
gaseous form is very unreactive, so can't be
used directly. Animals get what they need by
eating plants, which take their nitrogen in the
form of nitrates (NO_3) and other compounds
in the soil. But how do these compounds
get into the soil in the first place? The big
breakthrough in our understanding of these
processes was made by Dutch microbiologist
Martinus Beijerinck in 1885, when he isolated
nitrogen-fixing bacteria. These soil prokaryotes
take gaseous nitrogen from the air and
convert it into a form usable by plants. Plants
absorb it through their
roots and use it to make
amino acids, which are the
building blocks of proteins
and the raw materials for many other crucial compounds, such as
chlorophyll. When animals eat plants, the nitrogen compounds are
one of the nutrients they extract from this food. Nitrates are also
made by lightning. A bolt of lightning is strong enough to make
nitrogen react with oxygen in the air to form nitrogen oxides. These
dissolve in rain, forming nitrates, which end up in the soil.

Recycling system

Animals' feces and urine contain nitrogen nutrients. These are
broken down by decomposers, especially fungi and bacteria,
and nitrogen is returned to the soil as ammonia (NH_3).
Decomposers, including a wide range of invertebrate animals,
also recycle dead plants and animals, again returning nitrogen
compounds to the soil. Other bacteria in the soil then convert
these into useful nitrates. The level of nitrogen in the air is
constant because denitrifying bacteria convert nitrates back
into gaseous nitrogen, which returns to the atmosphere.

LEGUMES

Legumes (plants in the family Fabaceae,
including many kinds of beans) have
nodules on their roots filled with
nitrogen-fixing rhizobia bacteria. In return
for a place to live, these helpful germs
provide the plant with a constant supply
of nitrogen compounds. Farmers plant
legumes, such as clover, to add an extra
supply of nitrogen to a field's soil before
planting the next crop.

51 Chromosomes

CHROMOSOMES ARE THREADLIKE STRUCTURES IN THE NUCLEUS OF CELLS.
They are made from tightly coiled strands of DNA wound around proteins
called histones. Chromosomes carry the genetic code.

It was the German scientist Theodor Boveri who, in a series of experiments with
roundworms in the mid-1880s, showed that chromosomes are the vectors of heredity,
passing genetic information from generation to generation. DNA is a very long and
fragile chemical. Every human cell, for example, has about 2 m
(6.5 ft) of DNA inside of it—and that is a rather average amount.
All the DNA is carefully coiled up around histones to make
chromosomes, which are a protective scaffold for storing DNA.
Chromosomes got their name from an earlier term "chromatin,"
which referred to a tinted region seen inside cell nuclei. It was
later found that this tint was due to chromosomes, which were
at that stage in an unraveled, diffuse form. As a cell divides, the
chromosome takes on a compact and robust form, which was
what Boveri and others saw through their microscopes.

Organisms that reproduce sexually have diploid cells. That is,
there are two sets of chromosomes in the nucleus of most cells.
Only the gametes, or sex cells, have one set and are described
as haploid. One set of chromosomes comes from the female
parent and one from the male parent. Human diploid cells have
46 chromosomes in 23 pairs, common fruit flies have eight
chromosomes (four pairs), domestic cats have 38 (19 pairs), and
garden snails have 54 (27 pairs). The DNA in each chromosome
is divided into many sections called genes, which hold the code
for specific traits in the organism.

SEX DETERMINATION

The sex of a human is determined by two sex
chromosomes. The individual is female if these
sex chromosomes are the same (XX) and male
if they are different (X Y). Most animals use a
similar system. However, the sex of a crocodile
or turtle is determined by temperature. Eggs
in hotter or cooler parts of a crocodile nest are
female; males develop in medium temperatures.

52 Organelles

ANIMAL AND PLANT CELLS HAVE NUMBERS OF SUBUNITS within them, called organelles. They are surrounded by the cell's cytoplasm and are the site of specific metabolic functions.

All lifeforms can be divided into eukaryotes, which have a cell nucleus, and prokaryotes, which don't. Animals and plants, together with amoebae and similar single-celled organisms, are eukaryotes. As well as a nucleus, their cells have other membrane-bound units, or organelles. Prokaryotes do not have these clearly defined units. The nucleus contains most of the cell's genetic material, organized as DNA molecules coiled into chromosomes. Its primary function is to shield the DNA from the enzymes and other active chemicals at large in the rest of the cell. Similarly, other organelles provide microenvironments that are cut off from the rest of the cell, so offer the right conditions for a particular process.

Mitochondria are sausage-shaped organelles and are the power stations of cells. They release the energy from glucose and other foods that cells use to drive their many activities.

Mitochondria

Mitochondria are another type of organelle. They were discovered in 1890 by the German pathologist Richard Altmann, who called them "bioblasts." They are responsible for cellular respiration, the process where glucose is oxidized to carbon dioxide and water, with energy being released. These complex reactions largely take place on folded membranes within the mitochondria. The energy is stored in the form of adenosine triphosphate (ATP) and used for all the energy-consuming activities of the cell. There may be dozens of mitochondria in the most active cells, and several thousand in muscle fibers, which need a lot of energy. Mitochondria carry a small amount of their own DNA. The mitochondrial DNA (mtDNA) in sperm is not passed to the egg during fertilization, so all individuals always inherit their mtDNA from the mother.

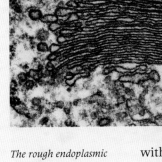

The rough endoplasmic reticulum organelle is a network of flattened bags and tubes that makes and transports proteins.

Other organelles

The endoplasmic reticulum (ER) is a network of tubular membranes. The "rough" ER is studded with small units called ribosomes, which make the proteins needed by the cell. The "smooth" ER manufactures lipids. The Golgi apparatus is a similar membranous structure. Its job is to package substances into membrane-bound vesicles for transporting out through the main cell membrane. Chloroplasts are found only in the green parts of plant cells. They are the sites of photosynthesis, where energy from sunlight is trapped by a green chemical called chlorophyll, which is bound to stacks of highly folded membranes. The number of chloroplasts in a cell ranges from one to about 100.

The function of a plant's chloroplasts is to conduct photosynthesis. Chlorophyll contained within the chloroplasts captures the energy from sunlight. This energy is then used to make sugar fuel from water and carbon dioxide. Oxygen gas is a waste product.

53 Orthogenesis

ORTHOGENESIS IS AN EVOLUTIONARY THEORY THAT SUGGESTS ORGANISMS have a tendency to evolve in a definite direction, most obviously by increasing in complexity. Very few modern biologists agree with its premise.

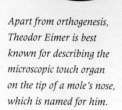

In the late 19th century many objections were raised to Charles Darwin's theory of evolution by natural selection. One concerned the relationship between variation in morphology and selection as a factor deciding the course of evolution. Darwin had assumed that the variations of individuals occurred more or less at random, so selection for advantageous characteristics was the main agent directing change in a population. But if variations tended to occur more readily in some directions than others—in other words, if they weren't random—then the direction of variation might itself control the course of evolution. If that were the case, selection would play only the negative role of eliminating those variations that were harmful. Several thinkers put forward variations on the theme of directed evolution. Some suggested mystical forces were at work, while others argued that there must be some kind of physical process involved.

Apart from orthogenesis, Theodor Eimer is best known for describing the microscopic touch organ on the tip of a mole's nose, which is named for him.

This traditional "Ascent of Man" view of the development of an upright-walking Homo sapiens from a knuckle-dragging chimpanzee is entirely orthogenetic. It suggests a direction to our evolution— from primitive to advanced. But the better way of understanding our place in nature is that our species is simply the last survivor of dozens of widely varied human-like species that evolved over the last 8 million years.

Trends not found

Wilhelm Haacke first introduced the idea of directed evolution in 1893, but it was popularized by fellow German zoologist Theodor Eimer. He wrote that when a whole species experiences changed environmental conditions, all the organisms would respond in the same way because they have the "same constitution." This idea went directly against Darwin's concept of natural selection. Eimer made the idea of "definitely directed variation" popular, calling it orthogenesis. He claimed there were trends in evolution with no adaptive significance that were therefore difficult to explain by natural selection. In 1898, he published *On Orthogenesis: And the Impotence of Natural Selection in Species Formation.* For a while the idea became popular with the public. However, the fossil record showed that long-extinct species had evolved in many directions from common ancestors, meaning natural selection has resulted in many different types of organisms—some successful and persistent, others less so.

54 Viruses

A VIRUS IS A TINY, INFECTIOUS AGENT THAT CAN REPLICATE ONLY IN THE LIVING CELLS OF OTHER ORGANISMS. Viruses can infect all kinds of life, including bacteria, as well as plants and animals.

One of the founders of virology, Martinus Beijerinck also discovered nitrogen fixation, a phenomenon that lies at the heart of life on Earth. He was an eccentric fellow who had a reputation for being rude to his colleagues. Despite his two great discoveries, he never won the Nobel Prize.

Only about 5,000 types of viruses have been described, although there are millions of varieties. Are they organisms or not? They have been described as "organisms at the edge of life," because while they carry genetic material, reproduce, and evolve through natural selection, they do not consume energy or have a cell structure. Viruses are tiny, typically one-hundredth the size of a bacterium, and have just two or three parts: Genetic material, either DNA or RNA; a protective protein coat called a capsid; and sometimes an outer fatty membrane. Viruses may have evolved from pieces of bacterial DNA that could move between cells, but now they are parasitic, hijacking the machinery of cells to copy themselves. When not infecting a cell, they are known as virions.

Viruses are the most abundant type of biological entity. A glass of seawater, for example, contains more viruses than there are people on Earth. The great majority of viruses are entirely harmless to people, with each one targeting a very specific set of hosts.

Discovery

In 1898 the Dutch microbiologist Martinus Beijerinck published the results of experiments showing that the tobacco mosaic disease, which damages tobacco crops, was caused by an infectious agent smaller than a bacterium. He called it a virus. His results accorded with the 1892 observations of Dmitri Ivanovsky, a Russian botanist. Beijerinck could not grow cultures of the virus, unlike bacteria, but he concluded that it could replicate and multiply in living plants. It wasn't until 1941 that X-ray crystallography conclusively proved the nature of the tobacco mosaic virus (pictured right).

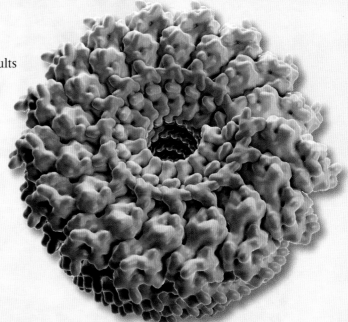

Attacking hosts

Viruses are transmitted in a variety of ways—in the air, in body fluids, and by direct contact. Viruses enter a cell and insert their DNA into that of the host. The viral DNA includes regulatory instructions that make the cell start producing copies of the viral DNA. Eventually the cell becomes so full it bursts, releasing the new viruses to infect other cells. If enough cells are impacted, the organism will become diseased. Generally, after a short illness the immune system can target and destroy viruses. However, there are a few human viruses, such as HIV, ebola, and West Nile virus, that debilitate the body so much that they may kill the host.

55 Succession

ENVIRONMENTS ARE ALWAYS CHANGING. SUCCESSION IS THE PROCESS of how living things take over a habitat to create an ecosystem. This process was first studied on windswept dunes in Indiana, USA, in the late 1800s.

Ecosystems never last forever. Eventually, a catastrophic event such as a forest fire or landslide (or human activity) will clear away the established species in a habitat, creating an empty gap. Nature reasserts itself after such events. For example, an abandoned lot in a city is first colonized by mosses, lichens, and small plants. These so-called pioneer species provide food for insects, which then attract birds and other animals. Later, shrubs and tree saplings sprout and grow. Increasingly complex communities of living things replace one another until a stable, unchanging community called a climax ecosystem is formed. This remains until the next catastrophe—on a small or large scale—creates a new gap.

When human activity is discounted, succession fills every habitat with a specific climax ecosystem. In eastern North America and Europe, for example, this is mostly broadleaf forest. Other regions with different climates produce different climax communities.

Succession was one of the first theories put forward by ecologists. Henry Chandler Cowles of the University of Chicago was among the first to study it, making observations of vegetation taking root in the Indiana Dunes in the late 19th century. In the 20th century, Frederic Clements identified intermediate stages in succession within an ecosystem, mapping a generic pattern of how gaps are always filled.

A tough pioneer plant sprouts from a lava field at the start of the long process of succession that will turn this lifeless habitat into a thriving ecosystem—until the next volcanic eruption.

Primary succession occurs when living things first colonize new habitats. This may happen after a disaster such as a volcanic eruption. For example, in 1963, a new island, later named Surtsey, rose from the sea off Iceland following an underwater eruption. Within just a few years, mosses, lichens, and then shore plants rooted in the ash and lava, and insects and, later, seabirds colonized the island. When a habitat is disrupted, for example through a forest fire or logging, the new growth is called secondary succession.

56 Genetics

THE SCIENCE OF INHERITANCE BEGAN WITH THE WORK OF GREGOR MENDEL in the 1860s, but it took until 1905 for the rest of the scientific community to give it a name.

THE GENE

A gene is a unit of inheritance, something that can be inherited and is not altered from generation to generation. However, the term is used in a confusing way. In everyday language it is most often used to describe a trait—the gene for hair color, for example. A better, more scientific definition would say a gene is a piece of DNA that codes for a single piece of protein.

MUTANTS

This albino cow carries a rare, mutant gene which stops it making pigments in the skin and hair. A mutation is a random change in the genetic code. Such a change can only be passed on to the next generation if it occurs in the germ line, or set of cells that make sperm or eggs.

The term "genetic" is derived from *genesis*, a Greek word that means "origin." Charles Darwin and others used "genetic" as an adjective to describe the as yet unknown mechanism of inheritance. However, using the word "genetics" for the scientific branch that studies heredity was the brainchild of William Bateson, an English biologist who rediscovered the work of Gregor Mendel in 1905. A few years later, the Dane Wilhelm Johannsen coined the term "gene" as the unit of inheritance. Other researchers were looking into DNA, a chemical found in chromosomes, but it took another 20 years before genes were firmly linked to DNA. After that, the science of genetics never looked back. Today, it is perhaps the most important field in biology, due to its strong links to medicine and technology.

Genetics uses other terms: "Allele" means a version of a gene. The gene for eye color, for example, has several alleles: Blue, brown, green, etc. The "genotype" refers to what alleles are inherited from the parents, while the "phenotype" is the final expression of those alleles. Basically, the work of a geneticist is to understand how a gene—or a sequence of DNA—creates a phenotype. That is no simple task. Genes do not determine a phenotype all by themselves. Instead, a phenotype is very often the product of the interaction between an individual's genes and its environment.

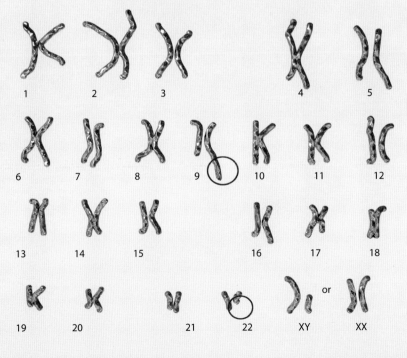

Eye color is a good example of phenotype. However, it does not follow simple Mendelian rules—few traits do. There are 16 genes involved, and the phenotype is not always the same in both eyes!

Geneticists focus a lot of research on abnormalities in human inheritance, such as when sections of chromosomes are missing, transferred (as seen here on the left from chromosome 22 to 9), or duplicated. As well as helping to solve health issues, the research also reveals how genes work in general.

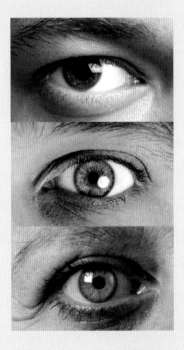

57 Cell Division

ALL LIFE IS MADE OF CELLS, AND ALL NEW CELLS ARE MADE FROM OLDER CELLS THAT HAVE DIVIDED. There are two kinds of cell division seen in nature: Mitosis and meiosis.

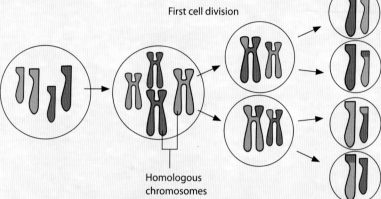

Second cell division

First cell division

Homologous chromosomes

Four daughter cells

The purpose of mitosis is growth, and that means making two genetically identical cells out of one. Microbiologists reported seeing cells divide in the 1870s, but the best description was given in the 1880s by the German Walther Flemming, who named the process "mitosis" from a Greek weaving term. Flemming chose this because he said that a dividing cell nucleus used a spindle of threads (now identified as microtubules made from proteins) that pulled the contents of the nucleus apart.

Each division follows a series of complex steps (see page 122 for more details), the first of which is when the chromosomes in the nucleus divide, forming two duplicates known as chromatids that are connected together. The chromosome then thickens to create the H-shape that we commonly associate with chromosomes. (Most of the time chromosomes don't look like this.) The spindle pulls the chromatids apart, making two sectors of chromosomes that move to opposing ends of the cell. These sectors will become the contents of the nuclei of the two daughter cells. A membrane forms, dividing the cell in half, with the organelles and cytoplasm roughly split between each half. The two cell halves eventually separate, creating two individual cells.

Body cells are diploid, which means they have a double set of chromosomes, one from each parent. Every chromosome carries a particular set of genes, so the double set of chromosomes can be organized into homologous pairs, which carry the same genes. Meiosis is the cell division that creates sex cells, such as sperm and eggs. Sex cells are haploid, meaning they have half the chromosomes of a normal cell. To halve the number of chromosomes, meiosis involves two cell divisions, resulting in four haploid daughter cells. In the first division, the spindle separates the homologous pairs. The second division then separates the chromatids, in a process very similar to mitosis. The sex cells are used in sexual reproduction: They will fuse together to make a zygote, which is the first cell of a new individual. In so doing, the gametes combine their chromosomes to make a diploid cell, which then grows by mitosis.

Meiosis is a cell division that creates four daughter cells with half the genetic material of the parent. There is a mixing process called recombination, which results in genetic material moving between chromosomes, so the four daughter cells are genetically unique.

Mitosis turns one cell into two identical daughter cells.

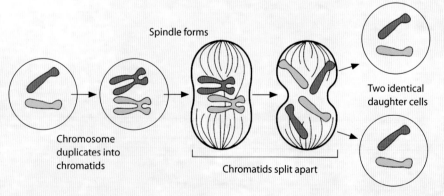

Spindle forms

Two identical daughter cells

Chromosome duplicates into chromatids

Chromatids split apart

58 Neurons

IN THE 1820S, CELL BIOLOGISTS DISCOVERED THAT THE BRAIN WAS MADE OF BRANCHING CELLS, which became known as neurons. It took another century to figure out the full complexities of their anatomy.

The ground-breaking development that allowed brain cells to be examined was the achromatic lens for microscopes. This could focus light of any color, which created the clearest views of brain tissue yet seen, and at much higher magnifications. The brain was sliced with a knife or teased into fine layers using tweezers. As the microscope was bought into focus, researchers could see the cells of the brain—at least the largest ones—for the first time. The brain is so densely packed with cells it was impossible to tell where one cell ended and another began. However, three anatomists, working independently, were able to make out the first views of the brain cells: Christian Gottfried Ehrenberg, Gabriel Valentin, and Jan Evangelista Purkinje. It is Purkinje's drawings that gave the best report of the discoveries, as he studied the large cells found in the cerebellum. His diagrams look a little like tadpoles or fish, with twisting, branching tails.

In 1872, the cell biologist Camillo Golgi (now remembered by the Golgi apparatus organelle in many cells) developed a dark dye that made it easier to study neurons. A few years later a German doctor, Bernhard von Gudden, invented a microtome, a device that could slice brain into ultrafine layers, making it possible to see individual neurons. However, no one could then agree on an interpretation of what they saw.

The cells appeared to have many "tails", not just one, around a central cell body, or soma. Most extensions were very short, and were named dendrites. However, one of them was thicker and much longer. This was initially known as the axial cylinder, but is now called the axon. Some researchers, including Golgi, believed that the axons and dendrites of neighboring cells were all connected together. That would make the brain—and the nerves that ran into the body—a vast single network. In the 1890s, the Spanish scientist Santiago Ramón y Cajal found that the cells were all separate from each other, and in 1897 the English biochemist Charles Sherrington showed there is no physical link between neurons. They communicate by an electric pulse along the axon, which then is transmitted to the next cell by chemical messengers leaping across a tiny gap called a synapse.

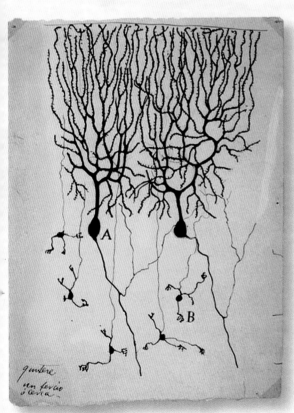

A hand-drawn diagram of a Purkinje cell in the cerebellum, named for its discoverer. This drawing was made by Santiago Ramón y Cajal, who helped figure out the full structure of the neuron in the 1890s.

Dendrites

Cell body

Axon

Synapse

Every neuron follows the same general structure. The nucleus is in the cell body, which is surrounded by hundreds of dendrites with a single axon. Axons carry messages from the cell, as electrical pulses, to the dendrites of its neighbors.

59 Pavlov's Dogs

PAVLOV'S FAMOUS EXPERIMENTS WITH DOGS were the first systematic study of animal learning. However, like many of the best discoveries, Pavlov's pioneering study came about by accident.

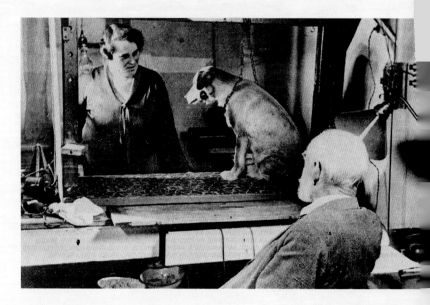

Russian physiologist Ivan Petrovich Pavlov did not start out as a behaviorist, but as an expert on digestion. In 1904 he won the Nobel Prize in Medicine for his work on the vagus nerve, which controls the flow of digestive juices to the stomach. Pavlov studied the digestive system of dogs, scientifically measuring the amount of saliva his subjects produced in response to food. One day he noticed that the dogs began to salivate before food actually appeared, when his assistant entered the room prior to feeding.

Pavlov focused on the salivary reflex in dogs because it was easy to quantify. Tubes collected the dogs' saliva so it could be measured.

RADICAL BEHAVIORISM

American psychologist B.F. Skinner was a pioneer of radical behaviorism. He analyzed animal behavior using a chamber, now known as a Skinner Box. An animal subject was placed inside, often a pigeon, where it was taught behaviors using food rewards or punishments from electric shocks. Skinner showed he could teach a pigeon to perform the same tasks as a monkey. Did that mean pigeons were as intelligent as monkeys? Skinner did not think so. Instead, he maintained that all behavior is determined by the consequences of previous actions and does not require a mental thought process. He even suggested human consciousness is a facade that only gives the impression of control over our actions.

Reflexes and conditioned responses

Pavlov reasoned that the dogs' impulse to salivate at the sight of food was instinctive, or "hard-wired" into the brain. He called this an unconditioned reflex. However, salivation at the sight of his assistant was a learned behavior or a conditioned reflex. The dogs had learned that the appearance of the researcher meant food would follow. Pavlov spent the next 20 years studying this response in dogs. In the classic experiment he used a bell as a neutral stimulus (one initially without a response). The dogs learned to salivate at the sound of the bell, which signaled food would follow. The food had to be produced quite quickly for the learned response to work. However, a conditioned reflex could also be unlearned. If the dogs learned that the bell no longer signaled food, they stopped salivating when it sounded. Pavlov also used a buzzer, a ticking metronome, and even electric shocks to condition his canine subjects. His research suggested that all behaviors were either innate reflexes or learned responses to reward or punishment.

Behaviorists study conditional behavior but acknowledge the role heredity plays in it. In contrast, radical behaviorists maintain that all behavior can be explained by conditioning.

60 Germination

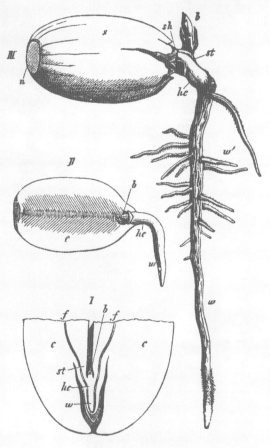

GERMINATION IS WHEN A SEED SPROUTS AND GROWS INTO A YOUNG PLANT. The processes behind this small miracle were not understood until the 20th century.

Angiosperms, or flowering plants, are the most familiar seed-producing plants. Their seeds develop inside the ovary, which develops into an outer covering called the fruit (it's not always something good to eat). However, gymnosperms such as conifers produce seeds inside cones.

The seeds of flowering plants have a tough outer coat, the testa. Inside is the embryo plus a food store, which may be contained in seed leaves called cotyledons or packed into tissue called endosperm. Angiosperms are classified according to whether they are monocots or dicots, bearing one or two cotyledons or "seed leaves." The embryo has an undeveloped shoot, the epicotyl, and an undeveloped root, the radicle.

All seeds require warmth, moisture, and oxygen to germinate. If all these are present, for example in tropical forests, seeds can germinate quickly. If not all are present, the seed remains dormant or inactive, for months or even years. Dormancy prevents seeds from germinating in fall, only to be killed by winter frosts. Some seeds require special conditions, such as cold temperatures or increased daylight to germinate, or the testa may need to be removed by passing through an animal gut.

When conditions are right the seed starts to germinate. It absorbs water and swells, so the outer coat splits. Stored food is converted into energy for growth. The radicle grows down into the soil as the first root. The plumule (shoot) grows upward taking the cotyledons with it. The young plant's roots start to absorb water and minerals, while the green parts begin to make food through photosynthesis.

The seed is a miniature life-support system, with a tough outer coat protecting the embryo and its food supply. The English naturalist Charles Darwin and his son Francis were among the first to investigate germination in the 1880s.

When light is to one side, auxin concentrated in cells on the opposite side of the plant stem stimulates growth toward the light.

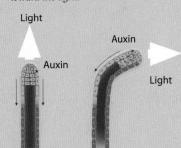

Light

Auxin

Auxin

Light

PHOTOTROPISM

Plants grow toward light to maximize photosynthesis. This is phototropism. A growth hormone called auxin is sensitive to light and is more concentrated on the dark side of the plant. Without auxin the cells on the bright side do not grow, but on the dark side, the cells lengthen rapidly. As a result the stem bends towards the light. This process ensures that the plant is always growing into sunny spaces. When a seed germinates in a dark place, auxin stimulates cell growth throughout the stem, and the shoot stays straight and grows fast, looking for light.

61 Model Organisms

MODEL ORGANISMS ARE SPECIES THAT ARE WIDELY USED IN LABORATORY RESEARCH because they are easy to breed and maintain. They are used to research animal behavior and genetics.

The nematode, or roundworm, Caenorhabditis elegans, is a model organism used to study embryology, such as the development of tissues and how genes control regulate it.

One of the most notable breakthroughs achieved using model organisms was Thomas Hunt Morgan's discovery, through his work on fruit flies, that chromosomes are the units passed on during inheritance. In the early 1900s, Morgan spent years breeding the fruit fly *Drosophila melanogaster* in jars in the famous "Fly Room" at Columbia University. He selected this fly for his research into genetics because it not only bred quickly, but frequently produced random variations called mutations, which allowed him to explore genetics in action. Morgan bred variants with different-colored eyes and wing shapes and looked for differences in their chromosomes.

The brown rat—which is not always brown—is a model mammal used for research into behavior, learning, and the effects of drugs.

Drosophila fruit flies are commonly used in research into genetics because they often manifest spontaneous changes called mutations. They also have extra-large chromosomes in their salivary glands which are easy to study.

How are model organisms chosen?

Yeast, bacteria, mustard cress, nematode worms, fruit flies, mice, rats, zebrafish, and guinea pigs are commonly used as model organisms in research. (The slang term "to be used as a guinea pig" means to be the first person to test something out—which is often the job of a model organism.) These species are easy to rear in labs, and also breed quickly, so that several generations can be studied in a short time. Research involving such species has helped us understand how heredity works, how cells grow and divide, and how living things store and use energy. Model organisms are also used to test drugs and new medical procedures.

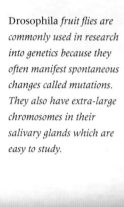

Model success

In the late 1800s, research with guinea pigs allowed Emil von Behring to develop an antitoxin for diphtheria. In the 1920s, Frederick Banting's research with dogs led to the development of insulin to treat diabetes in humans. Primates are sometimes used in laboratory tests because their physiology and behavioral responses are similar to humans. In the 1940s, Jonas Salk's work with rhesus monkeys led to the development of the polio vaccine, which saved millions of lives.

62 Biomes

EARTH'S SURFACE CAN BE DIVIDED INTO REGIONS THAT SHARE A SIMILAR CLIMATE, and as a result they have wildlife communities that must contend with a similar set of challenges. Regions of this kind are called biomes.

The term *biome* was first suggested in 1916 by American plant ecologist Frederic Clements. Professor of Botany at the University of Minnesota, Clements realized that biomes are not static, but gradually change over time. The plants that can grow in a biome are governed by climate, seasonal changes, terrain and altitude. In addition, each biome contains characteristic animals that are also adapted to these conditions.

Scientists may classify Earth's biomes in different ways, but they distinguish five main types of biomes: Aquatic, forest, grassland, desert, and tundra. Aquatic biomes include marine and freshwater ecosystems, such as lakes and rivers. Forests grow in areas that receive abundant rainfall. Forest biomes include tropical rainforests, which grow in torrid zones that are warm all year around, deciduous (broadleaved) woodland which appears in milder, more temperate regions, and the cool boreal forest, or taiga, of the far north. North of these coniferous forests stretch the treeless lowlands of the tundra. Similar climatic conditions, flora, and fauna exist on mountains, so scientists define two types of tundra: Polar and alpine tundra. Grasslands grow in regions that are too dry for forests to grow, but are wetter than deserts. There are two main types: Temperate grasslands, otherwise called steppes, pampas, or prairies; and tropical savannahs. Deserts are located in regions where sinking dry air prevents rainclouds from forming, so very little rain falls. They may be bordered by semidesert regions.

Temperate broadleaf forest
Coniferous forest
Grassland
Shrubland
Tropical rainforest
Desert
Mountain
Polar tundra
Aquatic

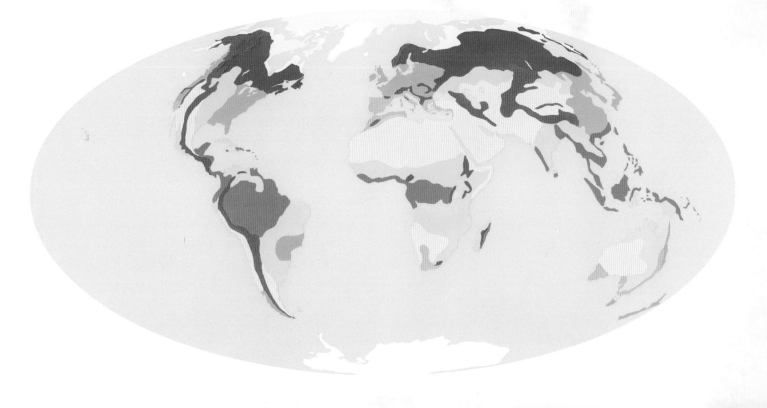

63 Cell Membrane

EVERY CELL, FROM A BACTERIUM TO AN OSTRICH EGG (WHOSE YOLK IS ONE *VERY* BIG CELL) is surrounded by an outer layer called the plasma membrane. This outer barrier and similar membranous structures in the cell are made from a double layer of oily chemicals called lipids.

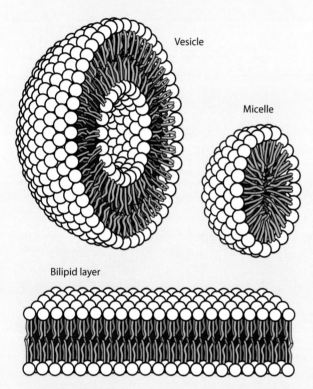

Vesicle

Micelle

Bilipid layer

Early cell biologists did not have the equipment to see the plasma membrane. Instead, they believed that cells were surrounded by walls. Plant and fungal cells (and bacteria) do have walls, but these are exterior structures that offer rigidity and protection. Animal cells do not have a wall. Wall or not, the environment of every cell, the tiny region where metabolism occurs, is protected within a membrane. The cell membrane was shown to exist in the late 19th century, and it was found to be semi-permeable. In other words, water and some materials could pass through it, but others could not. In 1925, its remarkable structure was revealed to be a double layer of lipids, which is the biochemical term for fats and oils.

Bilipid layer

A lipid molecule looks like a chemical jellyfish. The top is a glycerol unit from which three fatty acids hang. The membrane lipids also have a phosphate attached to the glycerol and this makes it hydrophilic, which means it can mix with water. The fatty acids are hydrophobic; they repel water and never mix with it. These two features combine to make a strong barrier a few billionths of a meter thick around the cell. The lipids form a double layer with the glycerols as the upper and lower surfaces. Inside the cell, they mix with the watery cytoplasm; on the outside they mix with any extracellular fluids. The inner region of the membrane is made up of the fatty acids, which mingle with each other, but do not let any water in.

As well as forming sheets inside and around a cell, membranes are also used to store material that must be kept separate from other chemicals in the cell. This material is held inside membranous spheres, or vesicles. These can also merge with the plasma membrane to release the contents into the cell. Fats and oils are transported in the blood as simpler micelles.

More complexity

With a more modern understanding, biologists today describe the cell membrane as being selectively permeable. Proteins embedded in the lipids act as pores and gates for letting certain chemicals in and out of the cell. The size difference between bacteria and other cells is also down to the cell membrane. Eukaryotic membranes have cholesterol mixed in which adds strength and allows them to form capsules hundreds of times bigger than a bacterial cell. We are rightly suspicious of cholesterol today for health reasons, but without it in our cells we would never have evolved beyond germs.

64 Food Webs

ANIMALS EAT OTHER LIVING THINGS TO GET ENERGY AND RAW MATERIALS FOR THEIR BODIES. In turn, they may be eaten by other animals. This flow of energy is shown in food webs.

All ecosystems contain many food chains, which interrelate to form a food web. Nowadays, food webs are a familiar idea, but the concept was first suggested in the 1920s by British zoologist Charles Elton. Elton pioneered the science of ecology by analyzing how living things interact with their environment. In his book *Animal Ecology* (1927), Elton used the term food cycle, which later ecologists called food webs.

Living things that produce their own fuel supply form the base of food chains. They are known as producers or autotrophs (which means self-feeders). By far the most common autotrophs are photosynthetic plants and algae, which use the energy in sunlight to transform carbon dioxide and water into glucose. (Some autotrophic bacteria get their energy from chemicals in rocks and hot-water vents on the sea floor. This process is called chemosynthesis).

This diagram shows part of a marine food web with many levels. Producers such as phytoplankton and algae sustain zooplankton, which are consumed by larger animals, and so on up to apex predators, such as sharks.

Up the chain

The plants are consumed by heterotrophs (other feeders); all animals and fungi are heterotrophs. Herbivores form the second level in the food chain, and are called primary consumers. They may fall prey to carnivores known as secondary consumers. These may in turn be eaten by more powerful carnivores, which are tertiary consumers. And so on up to apex predators at the top of the chain.

The chain does not end there. Decomposers (also called detritivores) are organisms that consume dead plant and animal matter. They include scavenging animals, fungi, and many bacteria, and their role is to recycle nutrients back into the soil or water, for plants to use. In this way, nutrients cycle through the food web, with the energy needed to process them being added by sunlight.

TROPHIC PYRAMID

Trophic pyramids are a way of analyzing the flow of energy through food chains. This diagram shows five levels in a grassland ecosystem. At each level, some energy is lost so less is available for the next level up. In most food chains, each level contains fewer organisms— there are fewer carnivores than herbivores, and apex predators are the scarcest of all.

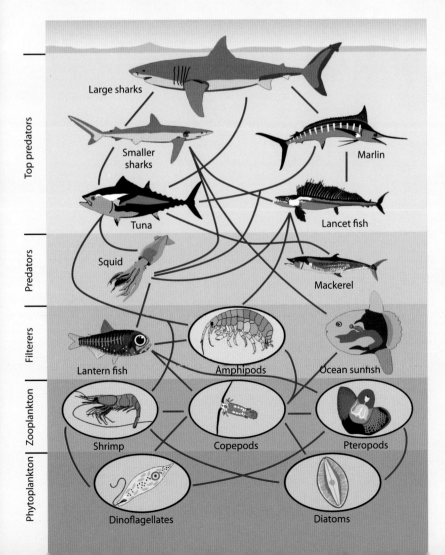

Top predators
- Large sharks
- Smaller sharks
- Marlin
- Tuna
- Lancet fish

Predators
- Squid
- Mackerel

Filterers
- Lantern fish
- Amphipods
- Ocean sunfish

Zooplankton
- Shrimp
- Copepods
- Pteropods

Phytoplankton
- Dinoflagellates
- Diatoms

65 Lysenkoism

LYSENKOISM WAS A POLITICAL MOVEMENT BORN IN THE SOVIET UNION IN THE LATE 1920S. Proponents of this theory rejected traditional scientific concepts such as genetics and Darwin's theory of natural selection.

The founder of Lysenkoism was an agricultural biologist named Trofim Lysenko. In the 1920s, he claimed to have developed revolutionary techniques that dramatically improved crop yields. At the time, the Soviet Union was running out of food and the Soviet authorities heralded Lysenko as the savior of the food crisis. Lysenko became a figure of the Soviet propaganda machine to denounce widely accepted theories, such as genetics and Darwinism. Lysenkoism remained unopposed in the Soviet Union until the 1960s, when his work was discredited by a new generation of scientists who followed mainstream scientific views.

66 Eusociality

SOME ANIMALS LIVE IN SOCIAL GROUPS IN WHICH ONE FEMALE PRODUCES ALL THE YOUNG, which are cared for by other adults in the group. This social system is called eusociality and is commonly found among ants, bees, and wasps.

Honeybees live in colonies in which one queen (A) mates with one or two drones (B). The rest of the colony consists of worker bees (C), which collect pollen with a hairlike basket on each back leg (D) and care for the larvae in hexagonal wax cells (E).

There are three main characteristics of eusocial animals. Adult females are grouped into reproductive queens, there is at least one breeding male (called king or drone), and a large number of workers. Overlapping generations of workers cooperate to forage for food and care for the young. The system works because the workers are closely related to the queen—often her daughters, and so are caring for their sisters, with whom they share many genes. Eusocial honeybees exhibit an unusual form of communication. Every day, workers leave the hive in search of sources of food. Upon their return, the bees perform an elaborate dance to reveal to other workers the direction and distance of the food source. Austrian biologist Karl von Frisch discovered this "waggle dance" in 1927 and won a Nobel Prize in 1973 for his work in deciphering its meaning.

NAKED MOLE RATS

The naked mole rat is one of only two mammals that exhibit eusociality. These rodents live in burrows beneath the grasslands of East Africa. Within the group, one female, called the queen, mates with a few breeding males. All the other adults in the colony are sterile and act as workers, foraging for food, maintaining the nest, and defending the colony from predators. (The other eusocial mammal is the hairy mole rat.)

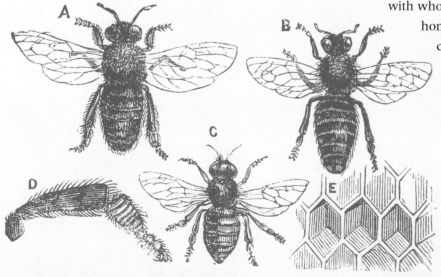

67 ATP

ATP, SHORT FOR ADENOSINE TRIPHOSPHATE, IS THE ENERGY CURRENCY OF THE CELL. This essential molecule is found in all living organisms and powers most of the processes that take place inside cells to keep those organisms alive.

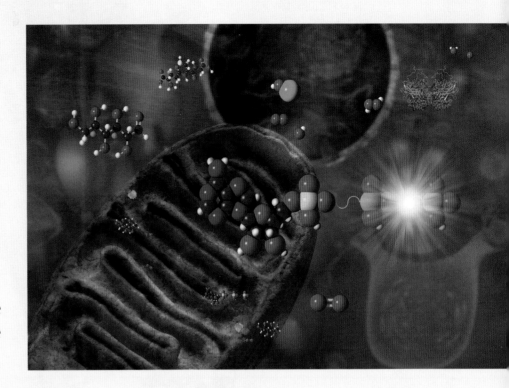

Every living organism needs a constant supply of energy to stay alive. Energy is used for processes such as cell division, the synthesis of proteins, and the transport of molecules through the cell. The cell harvests the energy it needs in a process called respiration (a slow, controlled oxidation of food molecules). The energy released by respiration is collected by ATP molecules, which travel to other parts of the cell, where the energy is then released again.

The production of energy-rich ATP molecules takes place inside organelles called mitochondria. These cellular "powerhouses" supply energy for cells to perform all their functions.

The structure of ATP

ATP was first discovered in 1929 by German biochemist Karl Lohmann and, independently, by Indian biochemist Yellapragada Subbarow and American biochemist Cyrus Fiske. A molecule of ATP consists of three main parts. A sugar called ribose forms the central part. A base called adenine (consisting of a ring of interlinked carbon and nitrogen atoms) is attached to one side of the ribose molecule. On the other side is a chain of three phosphate groups. The phosphate groups are the key to the energy-carrying ability of ATP.

How ATP works

ATP becomes active when it reacts with water (a reaction known as hydrolysis), producing a molecule of adenosine diphosphate (ADP) and a single phosphate group. The accompanying release of energy is given up to the metabolic processes that take place inside the cell. If an organism does not need immediate energy, the reverse reaction takes place, and the spare energy is used to reattach a phosphate group to ADP to form ATP. The oxidation of glucose as part of the citric acid cycle provides the energy for the conversion of ADP to ATP. Every glucose molecule produces about 30 ATPs. As a result, ATP acts like a battery, storing energy when it is not needed but releasing it instantly when the organism needs it.

68 Homeostasis

HOMEOSTASIS IS THE NAME GIVEN TO THE SET OF BIOLOGICAL PROCESSES that maintain and regulate a stable environment inside the body of a living organism. The word comes from the Greek, meaning "to stay the same."

KIDNEY

The kidneys are used for osmoregulation, the control of water (and dissolved salts) in the body. The kidneys act on instruction from hormones produced by the brain. The body takes in water from food and drink and loses water through the excretion of urine and feces (as well as sweat and vapor in the breath). When the water level in blood plasma is too low, the kidneys absorb more water back into the bloodstream and urine becomes more concentrated. When the water level in our bloodstream is too high, urine becomes more dilute. Water is removed from blood using a saclike filter, called Bowman's capsule, inside the kidneys.

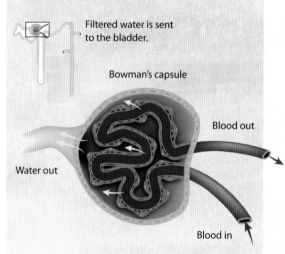

Filtered water is sent to the bladder.

Bowman's capsule

Blood out

Water out

Blood in

Homeostasis is a balance between input and output. Organisms take in elements from outside the body, such as air, water, and nutrients, and process them inside the body in strictly controlled conditions. Organisms then get rid of some elements, such as heat, water, and waste products, to preserve a balanced environment inside the body. French physiologist Claude Bernard was the first person to recognize the importance of this control system in the body, writing about it in a paper published in 1857. Many years later, in 1929, American physiologist Walter Bradford Cannon coined the term homeostasis.

Two main systems maintain homeostasis in the animal body. The autonomic nervous system provides an immediate response to changes in the environment. The endocrine system acts more slowly, using chemical messengers called hormones which travel around the body in the bloodstream. When the body senses a disruption to normal conditions, either one or both of these systems reacts to restore balance.

Temperature control

One of the most important homeostatic processes is thermoregulation—the maintenance of constant body temperature. The human body operates at a constant temperature of around 37°C (98.6°F). The hypothalamus in the brain is the body's thermostat. It measures the temperature of the blood flowing through it to detect any changes in body temperature. It also receives nerve impulses from temperature receptors in the skin. When the body gets too hot, blood vessels in the skin dilate. The body loses heat since blood then flows closer to the surface of the skin. At the same time, glands secrete sweat through the skin. The body cools as sweat evaporates. When the body gets too cool, blood vessels constrict to reduce blood flow to, and heat lost through, the skin. Shivering increases so that heat production inside the body also rises.

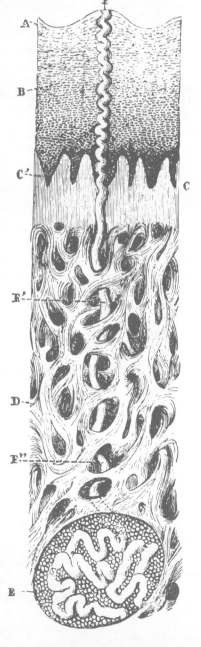

Glands beneath the skin produce sweat, which helps cool the body as it evaporates from the skin's surface.

69 Recombination

THE PROCESS OF RECOMBINATION DURING CELL DIVISION produces cells that have different genetic make-ups from each other and the parent cell.

Genetic recombination occurs during meiosis, the cell division that produces sex cells— sperm, pollen, eggs, etc. The idea of genetic recombination was first considered in 1931 by American geneticist Barbara McClintock and her student, Harriet Creighton. The two scientists were working with a strain of maize that had knob-like structures on only one of the chromosome pairs (called homologous chromosomes) inside its cells. McClintock and Creighton showed that this unusual structure altered the color and starchiness of the maize grain. McClintock and Creighton then tracked these traits by following the process of meiosis in later generations of maize plants.

During meiosis, parent cells divide and the resulting daughter cells, called gametes, are left with half the number of chromosomes as the parent. The halving begins when homologous pairs separate. The gametes of different organisms recombine during fertilization to form new individuals. McClintock and Creighton showed that parts of the maize chromosomes responsible for the identified traits were exchanged during meiosis. Evidence came from the fact that traits linked to the structures on the chromosome of the parent plant appeared in later generations of plants without the unusual chromosome structure, proving that genetic material had been exchanged.

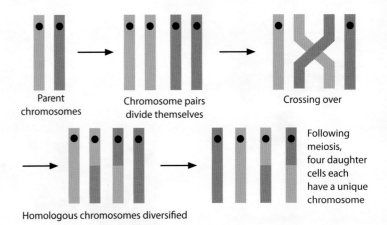

Parent chromosomes

Chromosome pairs divide themselves

Crossing over

Homologous chromosomes diversified

Following meiosis, four daughter cells each have a unique chromosome

Genetic recombination occurs during meiosis, when homologous chromosomes exchange segments of DNA ("crossing over") and later divide to form gametes, each with a different combination of genes.

Mechanism of recombination

The recombination of genetic material is a complex process guided by enzymes that cut the strands of DNA that make up the homologous chromosomes. During the first stage of meiosis, the homologous chromosomes duplicate themselves, line up, and strands of DNA from neighboring chromosomes cross over and exchange genetic material. Then the chromosome pairs split up and subsequently divide into four daughter cells (gametes), some of which may have a new combination of genes that differs from those of either parent. Genetic recombination is important because it boosts genetic diversity. This diversity allows populations to adapt to a changing environment. For many years, the work of McClintock and Creighton remained largely ignored by the scientific community. It was only vindicated through the advances in electron microscopy in the late 1930s, which allowed scientists to see recombination of chromosomes in greater detail, and the discovery of the structure of DNA in 1953 by British biologist Francis Crick and American biologist James Watson. In recognition of her contribution to genetics, Barbara McClintock received a 1983 Nobel Prize.

70 Vitamins

VITAMINS ARE SUBSTANCES THAT THE BODY NEEDS TO MAINTAIN GOOD HEALTH. Most cannot be made by the body and must be consumed in food.

One of the best-known vitamins is vitamin C, which is present in fruits and vegetables. This vitamin was first discovered in 1931 by Hungarian physiologist Albert Szent-Györgyi (1893–1986). His discovery was important in establishing the role of the citric acid cycle in cellular metabolism. All vitamins are essential to maintain good health, and people need 13 different vitamins to stay healthy. Vitamins A, D, E, and K are fat-soluble vitamins and can be stored in the body. They are found in foods rich in fats and oils, such as dairy products and oily fish. Vitamin A is essential for healthy eyesight, and without it, people can develop a condition known as night blindness, where they are unable to see in the dark. Vitamin D is required for healthy bones and teeth. The most important source of vitamin D is the Sun—the vitamin forms in the skin when exposed to natural sunlight. Vitamin D deficiency leads to rickets, which causes bone deformities. Vitamin E is needed to maintain blood cells. Vitamin K is vital for the process of blood clotting; it is produced by bacteria in the intestines as well as being present in food such as green vegetables. Vitamin C and a group of 12 vitamins called the B-complex are water-soluble. The body cannot store them, so people need a daily supply. The B-complex are required for vital processes such as cellular respiration and metabolism, as well as promoting the health of all the main body systems. B-complex vitamins are found in a range of foods, from cereals and dairy products to green vegetables and meat products such as liver.

Vitamins are often used as enzymes or as raw material for other helper chemicals used in metabolic pathways.

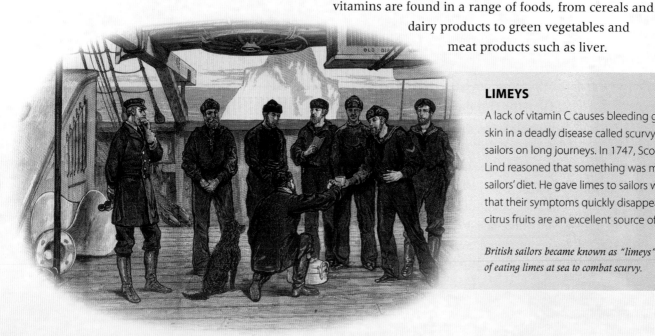

LIMEYS

A lack of vitamin C causes bleeding gums and bruised skin in a deadly disease called scurvy. It was a problem for sailors on long journeys. In 1747, Scottish surgeon James Lind reasoned that something was missing from the sailors' diet. He gave limes to sailors with scurvy and saw that their symptoms quickly disappeared. Limes and other citrus fruits are an excellent source of vitamin C.

British sailors became known as "limeys" because of the practice of eating limes at sea to combat scurvy.

71 RNA

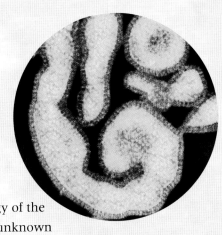

RIBONUCLEIC ACID (RNA) IS A CHEMICAL VERY SIMILAR TO DNA. It is used in the transfer of genetic information from the DNA in the nucleus to the cell at large.

The first step in the discovery of RNA and its role in the molecular biology of the cell came in 1869, when Swiss physician Friedrich Miescher isolated an unknown substance from the nuclei of pus cells. Miescher named this unknown substance "nuclein." A few years later, German biochemist Albrecht Kossel showed that nuclein was an acid containing five units, or "bases," namely adenine, cytosine, guanine, thymine, and uracil. Kossel won the 1910 Nobel Prize for Physiology or Medicine for his discovery of these "nucleic acids."

It took several more years before it was found that there were two types of nucleic acids—DNA and RNA. In 1933, Belgian biochemist Jean Brachet showed that DNA was localized in chromosomes, whereas RNA was present in the cytoplasm of all cells. Brachet also noted that cells rich in RNA tended to be those actively engaged in protein synthesis, which led him to propose that RNA plays a role in that process. The mechanism linking DNA, RNA, and protein synthesis, known as the Central Dogma of biology, came only after the discovery of the structure of DNA in 1953.

The influenza virus does not contain DNA; instead, it has RNA as its genetic material. The virus cannot replicate by itself, so it invades the cells of host organisms to do so, causing the symptoms of influenza.

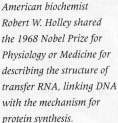

American biochemist Robert W. Holley shared the 1968 Nobel Prize for Physiology or Medicine for describing the structure of transfer RNA, linking DNA with the mechanism for protein synthesis.

Kinds of RNA

DNA is built from the "bases" adenine, cytosine, guanine, and thymine, and their sequence forms the genetic code. RNA's structure is similar to DNA's, but thymine is replaced with uracil. In addition, RNA is normally a single strand, while DNA is a helix made of two strands. Single-strand RNA—known as messenger RNA (mRNA)—carries genetic codes around the cell. The code is converted into specific proteins by structures called ribosomes. The ribosome is made partly of twisted RNA molecules (ribosomal RNA). The mRNA passes through the ribosome, and its base sequence is read using transfer RNA (tRNA), which builds up the protein step by step.

72 Ecosystems

AN ECOSYSTEM IS A COMPLEX OF ORGANISMS, TOGETHER WITH ALL THE NON-LIVING ELEMENTS in their environment. The study of the relationships within an ecosystem is the subject of ecology, a science born in the late 19th century, when German zoologist Ernst Haeckel adopted the term.

Alexander von Humboldt's intricate drawing of the plants he found on a volcanic island illustrates how changes in climate and other non-living factors create bands, or zones, of different ecosystems that form at different altitudes.

Haeckel was one of the first proponents of Darwin's new theory of evolution. He coined the term "ecology" to describe the study of the relationships between organisms and their environment, which was central to the theory of natural selection. British botanist Arthur Tansley was a pioneer of this new science of ecology. Drawing on the studies of other pro-Darwin scientists, such as Danish botanist Eugenius Warming, Tansley founded the British Ecological Society in 1913 and served as the first editor of the new *Journal of Ecology*.

Tansley introduced into biology the concept of an ecosystem in a paper he published much later in 1935. He described ecosystems as the basic units of nature and hoped to draw attention to the complex transfer of materials between living organisms and their non-living environment. Tansley's ideas quickly became popular and spawned a whole new field of scientific research, which continues to this day.

What is an ecosystem?

An ecosystem includes all the biotic, or living, things, such as animals, bacteria, and plants, in a given area, which interact with each other and with all the abiotic (non-living), elements, such as the climate, soil chemistry, and water supply. Within each ecosystem there are many individuals of the same species, which make up a population. The community includes all the populations of organisms of different species. Ecosystems are in equilibrium: If one population grows larger, it will have an effect on the other species, which will bring its population back down again. In this way every ecosystem exists in a delicate balance. All the biotic members of an ecosystem, together with all the abiotic factors, have their own niche and role to play, but they all depend on each other. Any new factor, such as the introduction of a new species, or a natural disaster, such as a fire or a flood, can potentially harm or destroy the ecosystem.

Human activities also contribute to the disruption of ecosystems. Many areas of rainforest have been cleared to make way for agriculture and human settlement, destroying the habitats of native animals and plants. Pollution releases harmful chemicals into the environment, causing widespread disruption to ecosystems either by destroying animal or plant populations or by contributing to eutrophication, which means enriching water bodies with extra nutrients that fuel excessive plant growth. In addition, invasive foreign species can decimate native species and upset the natural balance within an ecosystem. Today, conservationists seek to protect rare species by maintaining their ecosystems.

HYDROTHERMAL VENT

Hydrothermal vents are one of Earth's most unusual ecosystems. Deep beneath the surface of the ocean, super-hot water gushes out of cracks on the seabed. Heated by the volcanic activity under the ocean floor, these hydrothermal vents are home to many organisms that feed on chemicals in the heated water, including microscopic bacteria and giant tube worms.

73 Imprinting

IMPRINTING, OR THE PRINCIPLE OF ATTACHMENT, is a form of behavior in which young animals learn the characteristics of their parents, which are then said to be "imprinted" onto the young animals.

Konrad Lorenz feeds geese in a field near his home. Lorenz believed that animal behavior could only be studied in a natural environment and opposed studies in laboratories.

British biologist Douglas Spalding was one of the first people to study animal behavior, when he observed imprinting in chickens in the 1870s. His experiments were unknown until German biologist Oskar Heinroth discovered them in the early 20th century. Drawing on the work of both Spalding and Heinroth, in 1935 Austrian zoologist Konrad Lorenz published the first scientific study of imprinting, in greylag geese. Lorenz found that a newly hatched gosling instinctively bonded with the first moving thing it encountered. While most goslings would bond with their parents, the goslings in Lorenz's experiments could be made to imprint on Lorenz himself. Lorenz found that newly hatched geese imprinted within 16 hours after hatching. Lorenz also showed that goslings would imprint on inanimate objects, including a toy train moving around a train track.

With this work Lorenz helped start a new field of biology called ethology which studies animal behavior.

74 Citric Acid Cycle

THE CITRIC ACID CYCLE IS THE DRIVING FORCE BEHIND CELLULAR METABOLISM.
It is often called the Krebs cycle, in honor of its discoverer Hans Krebs. The cycle is the most significant stage in the breakdown of food to release energy for use by living cells.

As early as the 17th century, scientists started to realize the importance of chemistry in life. Oxygen, water, and carbon dioxide were all identified as being involved, but few predicted the sheer complexity revealed by a new breed of researchers called biochemists in the 20th century.

Earlier chemists had experimented with animals and plants to find out what chemicals living organisms needed to survive. In the 1630s, Jan Baptist van Helmont had identified the importance of water to plants (he thought it was their primary energy source). In the 1770s, Joseph Priestley and Jan Ingenhousz had shown that plants use up carbon dioxide in the atmosphere, producing "fresh air" in the process. This fresh air was oxygen, which is a by-product of photosynthesis, a process that uses solar energy to turn carbon dioxide and water into glucose sugar, the plant's fuel. Then in 1790, the French chemist Antoine Lavoisier revealed a reverse process. An animal breathes in oxygen and expels carbon dioxide. Combustion, the reaction that makes things burn, uses up oxygen and releases carbon dioxide, so Lavoisier's assumption was that the animal was using oxygen to burn fuel in the body. This process was termed respiration.

Plants use respiration as well to burn their glucose supply. At night when it is too dark for photosynthesis, they expel carbon dioxide just like animals do. Animals cannot make their own glucose, so they get their fuel by eating plants—or other animals that have eaten plants.

ANAEROBIC RESPIRATION

The citric acid cycle only occurs during aerobic respiration, when cells have a supply of oxygen, producing carbon dioxide and water along the way. In the absence of oxygen, cells can still produce energy in a process known as anaerobic respiration. The by-products of anaerobic respiration are either lactic acid or ethanol. Aerobic respiration is much more efficient than anaerobic respiration. For every molecule of glucose, anaerobic respiration produces only two molecules of ATP, while aerobic respiration produces up to 38 molecules of ATP. Athletes, such as the sprinter Usain Bolt, rely on anaerobic respiration when they run into oxygen debt during short bursts of intense activity. This causes a build-up of lactic acid, which contributes to the "burn" athletes feel in their muscles after exercise.

ACETYL-CoA

CITRIC ACID CYCLE

OXALO-ACETATE

CITRATE

MALATE

ISO-CITRATE

FUMARATE

a-KETO-GLUTARATE

SUCCINATE

SUCCINYL-CoA

Krebs showed that food is broken down in eight stages during the citric acid cycle, starting and ending with the formation of citrate from oxaloacetate.

Fuel from Food

Glucose is highly flammable, especially in pure oxygen. So why is it that living things do not explode as they "burn" all this sugar? Respiration releases energy in small steps in a series of biochemical reactions called the citric acid cycle. This cycle was first described by German-born British biochemist Hans Krebs in 1937. Krebs followed the metabolic pathway of glucose in minced pigeon breast muscle, since flight muscles have a high rate of respiration and are easy to experiment with. Before the cycle begins, the glucose is split into small units called acetyl groups (CH_3CO), which contain only two carbon atoms, along with hydrogen and oxygen. At the start of the citric acid cycle, the acetyl groups combine with a four-carbon compound called oxaloacetate to form a six-carbon compound called citrate (similar to citric acid, the sour stuff in lemon juice). In the following stages, this citrate molecule gradually breaks down with the help of enzymes. Eventually, the citric acid molecule is transformed into four-carbon oxaloacetate, which then restarts the cycle. Along the way, the reaction produces energy in the form of a compound called adenosine triphosphate (ATP), which is used to power other metabolic processes inside cells.

The citric acid cycle is an efficient process because only the acetyl groups derived from food are used up in the process. The enzymes that speed up each step, as well as the intermediate compounds that the enzymes act on, can be reused in successive cycles to produce more and more energy.

Krebs made his important discovery working in his laboratory at the University of Sheffield in England.

Prize Winner

It took Krebs five years of painstaking research to discover the stages of the citric acid cycle. Krebs owed much to the earlier work of Hungarian physiologist Albert Szent-Györgyi, who established the importance of some of the intermediary compounds in the cycle. Krebs also collaborated with German-born American biochemist Fritz Lipmann, who discovered coenzyme A (CoA), a key enzyme involved at the start of the citric acid cycle. In honor of their work, Krebs and Lipmann shared the 1953 Nobel Prize for Physiology or Medicine.

75 X-Ray Diffraction

IN 1952, THE BRITISH RESEARCHERS ROSALIND FRANKLIN AND RAYMOND GOSLING USED X-RAY DIFFRACTION to take "Photo 51." Some say this is the most important photo ever taken because it revealed the structure of DNA.

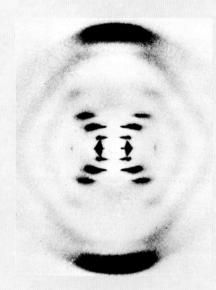

Waves of radiation scatter outward or diffract as they pass through small gaps, and in 1912, German scientist Max von Laue showed that X rays diffract as they travel through crystals. That provided a means of measuring the spaces between atoms in the crystal, and the angles at which the X rays emerged could be used to suggest possible structures for the molecules. By the 1950s, X-ray diffraction was being used to investigate complex chemicals found in cells, such as vitamins and cholesterol. Franklin and Gosling pulled DNA into strands and shone a beam of X rays at it for more than 60 hours. The X rays scattered as they passed through the DNA and produced an image on a film, with the shape of the image giving clues as to the structure of the molecule. The structure of DNA was the biggest prize in biochemistry, and Photo 51 was instrumental in revealing its structure in 1953.

The darker patches on Photo 51 reveal where X rays, diffracted by repeating features within the DNA molecule, have bombarded the film. The dark patches at the top and bottom represent DNA bases, which together make up the genetic code.

76 Action Potential

IT HAD BEEN KNOWN SINCE THE LATE 18TH CENTURY that electricity was a factor in energizing nerves, but the exact mechanism by which a nerve cell could produce an electrical signal remained a mystery.

The Italian physician and biologist Luigi Galvani had demonstrated in the 1790s how nerves and muscles were stimulated by some kind of electricity. Researchers found that electrifying the brain led to all kinds of effects, such as involuntary facial expressions. In 1875, Richard Caton recorded electrical fields being made by the brains of monkeys and rabbits. But how were the electrical impulses made?

Giant nerve fiber

The breakthrough came when two English researchers, Andrew Huxley and Alan Hodgkin, began to study the nerves of squid. This animal was chosen because of its unusually large axon. Although only about 0.5 mm (0.019 in) in diameter, this nerve fiber runs the length of a squid's body. The two scientists studied this giant nerve using a technique called voltage clamping. They changed the voltage of the axon and recorded how that varied the way charged particles called ions moved in and out of the nerve. Huxley and Hodgkin began their research in the 1930s but were interrupted by World War II. In 1952 they were able to present their results. They revealed that a neuron or nerve cell did nothing most of the time, but when the time came to send a signal it used charged potassium, sodium, and chloride ions to create a surge of electrical potential that traveled along the axon.

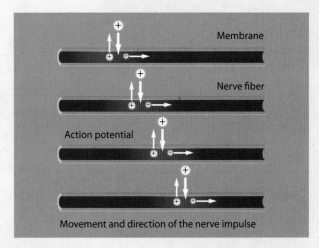

Movement and direction of the nerve impulse

Resting potential

A neuron is described as "at rest" when it is not sending a signal. At rest, it has a negative charge on the inside and a positive charge on the outside because some charged ions can move across the semipermeable cell membrane and others cannot. Negatively charged chloride ions are blocked from leaving the cell, thus creating a pool of negative charge inside. Positively charged potassium ions are free to move through the membrane in either direction, so they can flood into the cell to even out the difference in charge either side. But, in addition, the nerve axon pumps out positively charged sodium ions faster than the positively charged potassium ions can get in. All this movement requires energy, and partly explains why the human brain requires 20 percent of the body's fuel supply. It needs feeding even when it is apparently doing very little. The difference in voltage between the inside and outside of the axon is called the resting potential, and it is about −70 millivolts (mV).

The action potential travels along the axon at about 150 meters per second (about 500 feet per second).

Chemical stimulus

If the axon is instructed to send a signal current, a chemical stimulus from the neuron causes channels to open in the axon's membrane close to the main part of the cell. When this happens, the channels allow positively charged sodium ions to flood back into the cell, changing the potential in this part of the axon. If it gets above the critical level of −55 mV, the process accelerates and more sodium channels open. The outside of the axon then becomes negative and the inside positive; the polarity of the axon has reversed. The system then switches back to "normal," with the sodium ions pumped out of the axon and the resting potential restored. However, the change in polarity moves along the axon, repeating the process over and over to create the spike of "action potential." The spike is the nerve impulse, and the action potential of a motor nerve is what causes the muscle fibers to contract.

A portrait of Alan Hodgkin poses him in his laboratory in 1963, the year he received the Nobel Prize, along with Huxley and a third colleague called John Eccles.

77 Double Helix

DISCOVERING THE STRUCTURE OF DNA was one of the most complex yet revealing puzzles tackled by biochemistry. Once achieved, it spawned a whole new science.

ROSALIND FRANKLIN

An English chemist and X-ray crystallographer, Franklin is best known for her work on the structure of DNA. Her X-ray diffraction images included Photo 51, which indicated the molecule's double-helix structure and led to a flowering of the science of genetics. Sadly, Franklin died before her important role in DNA research was recognized.

By 1928, it was becoming clear to the scientific community that DNA was the genetic material that Gregor Mendel had proposed. The big challenge for biologists now was to determine how DNA's constituents were able to pass on genetic information.

Competing teams

By the early 1950s, several research teams were involved in the search for the structure of DNA. It was painstaking work to map such a large and complex molecule, and the most reliable approach was to use X-ray crystallography to calculate DNA's geometry. At the Cavendish Institute in Cambridge, Englishman Francis Crick, a physicist who had turned to biology after the war, had teamed up with American James Watson to model the structure of DNA from whatever evidence was available. Watson and Crick were convinced they had found the answer when they proposed a three-stranded helix, with the nucleic acids, known as bases, on the outside of the molecule and the negatively charged phosphate groups on the inside. There was a problem though: So many negatively charged phosphate ions in the core would repel each other and blow the molecule apart. They suggested positive calcium or magnesium ions might balance the charges, and asked Maurice Wilkins and his assistant Rosalind Franklin, from London's King's College, to look at their structure. Franklin highlighted a major problem: Any positive ions in the core of the DNA would be surrounded by water and that would stop them canceling the destructive negative charges.

The final breakthrough

In May 1952 Rosalind Franklin used X-ray diffraction to image DNA. One of these images, the famous Photo 51, indicated

James Watson and Francis Crick (below), along with Maurice Wilkins, were awarded the Nobel Prize in Physiology or Medicine in 1962 for their work on uncovering the structure of DNA.

The DNA molecule is now known to be similar to a ladder coiled into a double helix, with rungs of paired bases linked by hydrogen bonds.

that the DNA was two strands arranged in a spiral—a double helix. Maurice Wilkins showed the photograph to James Watson. He and Crick used it to construct a new model of DNA, which they announced the following year.

The two sides of the ladder-shaped helix were made from ribose, connected together with phosphates. Pairs of bases made up the "rungs" linking the sides. The four bases always bonded with a specific partner: Cytosine partnered guanine, while thymine linked to adenine. The model produced by Watson and Crick spelled out a code using cytosine, guanine, thymine, and adenine as characters, generally represented with the letters CGTA. A gene was a strand of DNA with a unique sequence of bases. This discovery laid the basis for the still-developing science of genetics, but many questions remained. Despite knowing a lot more about how the DNA code is passed on and functions, the mechanisms that lead from a genetic code to a physical trait are more complex.

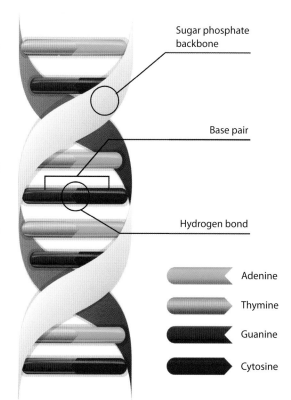

Sugar phosphate backbone

Base pair

Hydrogen bond

Adenine

Thymine

Guanine

Cytosine

78 Ecological Niche

AN ECOLOGICAL NICHE IS THE POSITION A PLANT, ANIMAL, OR OTHER ORGANISM has in its environment together, with its role there.

Working at Yale University, the British-born zoologist G. Evelyn Hutchinson, sometimes called the "father of ecology," was keen to adopt a rigorous, scientific approach to ecology by examining all the processes at work in biological systems or ecosystems, whether they are physical, chemical, or geological. In 1957 he described an ecological niche as an "n-dimensional hypervolume," where the dimensions are the environmental factors that enable a species to survive. He suggested that an organism's role in its niche includes how it feeds, reproduces, and finds shelter, and also how it interacts with other lifeforms and the physical environment. Hutchinson's work inspired others to explore and explain the variety of resources used by a single species (niche breadth), the ways in which competing species use the environment differently so they can coexist (niche partitioning), and the overlap of resource use by different species (niche overlap).

An oxpecker picks ticks from the fur of a buffalo. The requirements of the two species are different but they share the same habitat.

79 Biophysics

BIOPHYSICS APPLIES THE METHODS OF PHYSICS TO STUDY biological systems. The very latest advances in biology are almost unthinkable without a large input from physics.

Adding an understanding of atomic physics to biology allows scientists to understand how complex proteins, such as this adenovirus, respond to other chemicals.

There has always been an overlap between biology and physics, but the term "biophysics" wasn't used until 1892, and the discipline was properly recognized only in 1958 when the Biophysical Society was established in the USA. Biophysicists work at all levels, from the atomic and molecular, to the organization and behavior of cells, organisms, and environments. They use techniques developed to understand physical phenomena to answer questions that are hard for biologists to otherwise solve. For example, the structure of DNA was finally figured out using X-ray technology pioneered by physicists. Biophysicists today study questions such as how quantum physics is working in metabolism, especially in photosynthesis; how proteins can work like "machines" in cells—and could be used elsewhere; and how networks of nerves use a code to send messages.

80 Central Dogma

THE CENTRAL DOGMA IS AN EXPLANATION OF HOW GENETIC INFORMATION flows within an organism. It was first stated in 1958 by Francis Crick, one of the scientists who had discovered the structure of DNA.

KINDS OF RNA

There are many types of RNA in a cell but the Central Dogma relies on four. Messenger RNA (mRNA) encodes the amino acid sequence of a protein. Transfer RNA (tRNA) brings amino acids to the ribosome organelles during translation. Ribosomal RNA (rRNA), with ribosomal proteins, makes up the ribosomes. And finally, small nuclear RNA (snRNA; also called "snurps") forms complexes with proteins that are used in RNA processing; snRNA is found only in eukaryotic organisms.

Francis Crick, one of the discoverers of DNA's helical structure, had shown that genetic information was written in the four-letter code made by sequences of the four nucleotide bases: Adenine (A), thymine (T), cytosine (C), and guanine (G). The DNA is the store of information, but information about what? In 1955 it was found that genetic code is the instructions for building proteins, which are used by the cell as enzymes or other functional chemicals. Proteins are

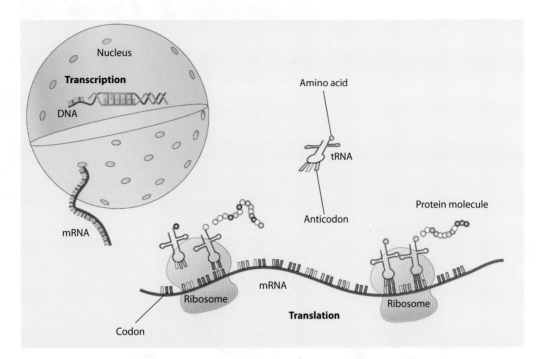

The Central Dogma is a multistage process that transfers information from the DNA in the nucleus to protein factories, called ribosomes.

complex polymers made up of long chains of several dozen amino acids. Each gene's code corresponds to a unique sequence of amino acids. Crick characterized the process of turning genes into proteins as a dogma to illustrate the fact that genetic information only ever moves in one direction: Information stored in the DNA code is transmitted to the cell using RNA, which builds or synthesizes a specific protein. However, the structure of a protein is never used to encode the DNA. In addition, the Central Dogma applies to all forms of life, using a similar code and the same translation system.

Follow the process

The information transfer begins with transcription. This is when the DNA helix unravels into two separate strands inside the cell's nucleus. Only one of these strands carries a gene—the opposite strand is a non-coding mirror image of it. This non-coding, or "anti-sense" strand is used as a template to make a corresponding strand of mRNA (messenger RNA), which is a copy of the coding, or "sense," DNA strand.

The mRNA then leaves the nucleus and moves to a ribosome, where it is translated into a protein. The ribosome is made from two units, themselves structured around folded RNA molecules. The mRNA is threaded through these ribosomal units. It moves along three bases at a time. Each three-base unit makes a codon, and each codon relates to a specific amino acid in the protein. As the mRNA passes through the ribosome, the codons are read by an anticodon of three bases in a transfer RNA (tRNA) molecule. The tRNA's role is to haul the next amino acid into the chain as the protein is built step by step.

CODONS

A codon is a sequence of three nucleotides that corresponds with a specific amino acid. There are also codons for "start" and "stop," signaling when a protein chain begins and ends during protein synthesis. DNA and RNA molecules are written in a language of four nucleotides. There are 64 possible codons, most of which code for one of the 20 amino acids used in proteins. There is more than one codon for each acid.

Codons form the key between the four-letter language of DNA and the 20-letter language of proteins. The idea of a triplet code was investigated by the nuclear physicist George Gamow (left), more famous for the Big Bang Theory. Others refined Gammow's idea to unlock the genetic code.

81 Animal Behavior

THE SCIENTIFIC STUDY OF ANIMAL BEHAVIOR in the wild is called ethology. It seeks to explain how different behaviors benefit individuals or communities of animals.

Animal behavior has fascinated zoologists for centuries, but the birth of the modern discipline of ethology is generally understood to have taken place in the 1930s with the work of Dutch biologist Nikolaas Tinbergen and Austrian biologists Konrad Lorenz and Karl von Frisch, all three of whom went on to win the Nobel Prize. Tinbergen's 1963 paper *On Aims and Methods in Ethology* asked four questions that he argued need to be answered to understand any behavior: What stimuli produce the behavior? How does the behavior contribute to the animal's success? How does the behavior develop during the animal's life? How did the behavior first arise in the species?

Jane Goodall studied chimpanzee behavior at Gombe Stream National Park, Tanzania, from 1960 to 1975. She later set up the Jane Goodall Institute to continue the investigation of these apes' behavior.

Nikolaas Tinbergen's 1963 paper On Aims and Methods in Ethology *established ethology as a serious scientific discipline.*

Chimps, gulls, and ducklings

Jane Goodall is probably the best-known animal behaviorist. Her detailed 55-year study of the social interactions of chimpanzees in Tanzania completely revised our understanding of these primates. She discovered that they have complex social behaviors, famously observing that they have distinct personalities and they hug and tickle each other. Perhaps her most remarkable observations were of chimps not only using, but also making tools. Goodall watched one chimp poke stalks of grass in the holes of a termite mound, then remove the stalks from the hole covered with clinging termites; effectively "fishing" for the insects. On other occasions, she watched chimps take twigs from trees and strip off the leaves to make them more effective, clearly a form of toolmaking. Nikolaas Tinbergen studied how herring gull chicks instinctively peck at red spots on their parents' beaks to encourage them to regurgitate food. Konrad Lorenz is most famous for his work on imprinting, the phenomenon by which young animals, such as ducklings, often copy their parents.

82 Cladistics

CLADISTICS IS THE METHOD OF CLASSIFYING ANIMALS AND PLANTS based on shared characteristics that are traceable to a group's most recent common ancestor, but which aren't present in more distant ancestors.

A group that shares characteristics can be deduced to have a common ancestral history and therefore the members are relatively closely related. What is now called the cladistic method has its roots in the work of zoologists such as Peter Chalmers Mitchell in the early years of the 20th century. The German Willi Hennig explained the idea more fully in a book published in 1950, but it wasn't until 1966, when the book was translated into English, that the method became more generally accepted by taxonomists internationally. It still wasn't accepted by everyone, and until at least the end of the 1970s it competed for popularity with phenetics, which based classification on morphology—what the organisms look like—without examining ancestry. Phenetic analyses can become misled by the process of convergence, which may give rise to similar-looking but not closely related organisms, for example pronghorns in North America and true antelopes in the Old World. Cladistics is better at (although not immune from) avoiding this problem. The analysis of organisms' genetic make-up can now be used to help construct accurate tree diagrams of relationships, or cladograms.

Cladograms are diagrams that depict the relationships between different lifeforms called "clades." By showing these relationships, cladograms reconstruct the evolutionary history, or phylogeny, of the group of organisms. This diagram shows the phylogeny of animals, showing three major clades: The Deuterostomia, which includes vertebrates; the Ecdysozoa, which covers crabs, insects, and many other bugs; and Lophotrochozoa, which includes the mollusks.

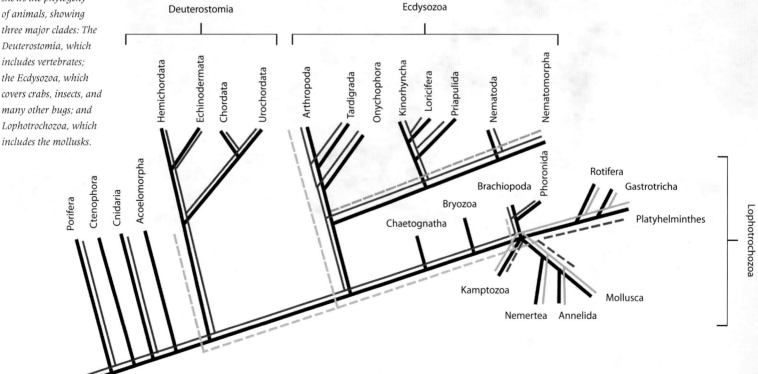

83 Symbiogenesis

EVIDENCE FROM DNA TELLS US THAT THE FIRST LIFEFORMS were prokaryotes, such as bacteria and archaea, with relatively small, simple cells. The theory of symbiogenesis suggests that the larger, complex cells of protists and multicellular organisms arose when prokaryotes began living as a team.

Symbiosis occurs when two different species benefit from living together. In 1967, the American geneticist Lynn Margulis set out an amazing theory that eukaryotic cells, with their complex internal structures, evolved from unrelated prokaryotes that began living together. The evidence comes from the mitochondria and chloroplasts in eukaryotic cells, which were noted to resemble bacteria as long ago as 1910. Later discoveries showed that these organelles carry their own DNA. Analysis of that DNA reveals that the mitochondria found in almost every living cell have DNA similar to proteobacteria, also known as purple bacteria. This is a large group of prokaryotes, some of which cause diseases, while others are nitrogen-fixing bacteria in soils. The chloroplasts in plant cells, meanwhile, are closely related to cyanobacteria, also called blue-green algae. These were the first photosynthetic organisms.

One-off event

Margulis's idea is known as symbiogenesis, or endosymbiotic theory, because the constituents of the cell are now dependent on each other. However, that may not have been how it started. The intruding bugs may have been food engulfed by a larger cell, or may have entered as a parasite. The nuclear DNA in eukaryotes is closer to that of archaea than that of bacteria, so it is assumed that the cell that formed the haven for the endosymbionts was an archaea that had evolved a large and convoluted cell membrane.

The first eukaryotic cell appeared at least 1.5 billion years ago, and whatever the process was, it is thought to have occurred successfully only once. All protists, fungi, plants, and animals are descended from that one cell. It is thought that the first eukaryote fed on other organisms, and did not use photosynthesis. All fungi and animals evolved from this. The photosynthetic chloroplast was a later addition, which gave rise to plants.

Nucleoid (containing DNA)
Prokaryotic cell
Cytoplasm
Cell membrane
1
Cell membrane infoldings
Extinct descendants
† 2
Nucleus
Endomembrane system
Nuclear membrane
Endoplasmic reticulum
†
3
Proteobacterium
First eukaryote
†
4
Mitochondria
Cyanobacterium
Mitochondrion
Ancestor of animals, funghi, and other heterotrophs
5
Chloroplasts
Ancestor of plants and algae

A possible route for symbiogenesis involves these steps: 1) A prokaryotic cell, probably an archaea, grows larger and develops folds in its membrane that increase its surface area. 2) The folds break off from the cell membrane, forming membranes around the genetic material (the nucleus) within the cell and creating the endoplasmic reticulum membrane system. 3) An oxygen-using purple bacteria enters and avoids destruction. It can reproduce and is passed on during cell division. 4) The bacteria uses oxygen to release energy for the cell as a primitive mitochondrion. 5) A cyanobacteria enters the cell and begins providing sugar via photosynthesis, creating the first plant-like cell.

84 Kin Selection

THE "SELFISH GENE" IS A FAMILIAR PHRASE IN POPULAR ACCOUNTS OF BIOLOGY.

It does not mean that all organisms are selfish. In fact, it was coined to explain why animals sometimes help each other.

Sometimes, animals behave in a way that increases the fitness (ability to survive) of other individuals but decreases their own chances of reproductive success. For example, lionesses sometimes nurse cubs that are not their own, and worker honeybees defend their colony by making suicidal attacks on intruders. In the late 1960s, British biologist William D. Hamilton devised the concept of genetic relatedness to explain such altruistic behaviors. He arguedw that apparent altruism led to the success of genes shared by related animals. This concept is called kin selection.

In the family

Hamilton devised a formula for when altruism might be seen: r × B > C, where r is the genetic relatedness between the two animals (the proportion of genes they share), B is the benefit (the number of offspring produced) gained by the recipient of the altruism, and C is the cost (the number of offspring not produced) suffered by the altruist. Hamilton argued that if a potential provider of help (the altruist) can more than make up for losing its offspring by adding to the offspring of the related beneficiary, then altruism is seen in a population. The evolutionary biologist Richard Dawkins popularized this view of natural selection in his 1976 book *The Selfish Gene*. The book outlines natural selection in terms of the gene, saying that organisms are simply vehicles for propagating genes. Just as all anatomical traits aid in that goal, so do behaviors, including altruistic ones.

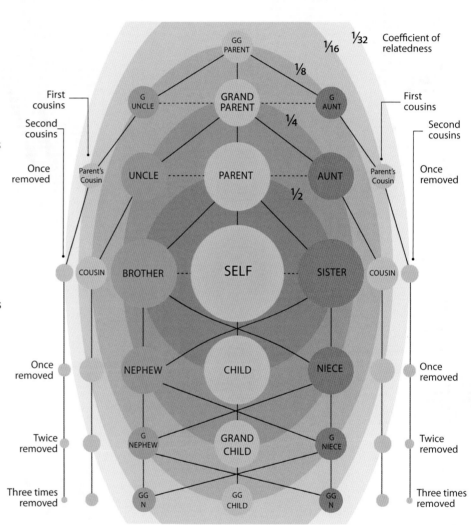

This chart shows the fraction of genes you share with your family members. Third cousins are no more related to you than a complete stranger.

85 Punctuated Equilibrium

THIS THEORY PROPOSES THAT ORGANISMS EVOLVE IN OCCASIONAL PHASES of rapid changes that take place in part of its population. The result is a new species branching away from its direct ancestor.

Punctuated equilibrium is usually contrasted with the theory of phyletic gradualism, which maintains that evolution occurs through the gradual transformation of whole lineages. As a result of their examinations of the fossil record, the paleontologists Niles Eldredge and Stephen Jay Gould put forward the theory of punctuated equilibrium in 1972. They argued that species tended to appear suddenly in rock strata, and then remain very similar—in stasis. When significant change did arise it was restricted to relatively rapid events, leading to new species splitting from a parent species. Some people misunderstood the theory to mean that a sudden change happened between one generation and the next, but that was not what was being proposed—rather, that rapid bursts of evolution could occur over the course of, say, a few tens of thousands of years, which is fast in the context of the geological timescale.

Phyletic gradualism

Morphology

Time

Punctuated equilibrium

Varying pace of change

The idea that the pace of evolution varies was not a new one. Darwin had acknowledged that some species evolved faster than others. Later, Ernst Mayr wrote in the 1950s of "genetic revolutions" occurring when a small population of a species becomes isolated and accumulated changes arising from genetic mutation become dominant—if they are advantageous. But Gould and Eldredge took the idea a step farther, arguing that gradual evolutionary change was rare in the geological record, and once a species had become established it tended to remain in stasis. While not all paleontologists support the theory, there is no denying that stasis is common in the geological record, including extreme cases such as the fern *Osmunda claytoniana*, whose form has been unchanged for 180 million years, and the *Lingula* genus of brachiopods, which has remained virtually identical for 500 million years.

The traditional idea of phyletic gradualism envisages species gradually evolving (angled lines) and slowly branching into new species. Punctuated equilibrium argues that species generally remained almost unchanged (straight vertical lines) with new species arising through evolution only occasionally (horizontal branches).

RED QUEEN HYPOTHESIS

Proposed by Leigh Van Valen in 1973, this theory references a line by the Red Queen in Lewis Carroll's *Through the Looking-Glass*, when the Queen tells Alice: "Now, here, you see, it takes all the running you can do to keep in the same place." The hypothesis proposes that organisms must constantly adapt and evolve simply to survive when in competition with ever-adapting rivals; if they don't they will go extinct. The hypothesis has also been described as "the evolutionary arms race."

86 Genetic Engineering

THE MANIPULATION OF AN ORGANISM'S GENOME USING BIOTECHNOLOGY IS **CALLED GENETIC ENGINEERING.** It includes the transfer of genes within and across species boundaries to produce improved or different organisms.

People have altered animal genomes for thousands of years using selective breeding, but this is a slow and inexact process. Once scientists understood the role of DNA, they began looking at ways of manipulating it. In 1972 Paul Berg combined the DNA from a monkey virus with that from a lambda virus to create a new "transgenic" virus. Two years later, Rudolf Jaenisch created transgenic mice by introducing DNA of a retrovirus (that can affect the host's DNA) into their embryo cells. These were the first genetically engineered mammals. Scientists also considered how to make crops more resistant to frost and disease. The first field trials of genetically engineered plants took place in 1986; these were herbicide-resistant tobacco plants. Despite controversy, by 2009 there were 11 transgenic crops being grown commercially in 25 countries.

The process involves isolating the gene to be inserted. Then it can either be inserted by using a virus, which enters a target cell and introduces the DNA, or a gene gun can be used to blast cells with vast numbers of the DNA bound around tiny particles of gold. Most of the DNA and cells are destroyed, but in a few cases, the DNA enters a cell intact and is incorporated into the material of the cell's nucleus.

Rudolf Jaenisch is a pioneer of transgenic science. He created the first transgenic mammal and has since used genetically engineered mice to research the fight against neurological disease and cancer.

Scientists have used a jellyfish gene as a "marker" to show whether gene transfer to mice embryos has been successful. The gene produces a protein that gives jellyfish a green fluorescence. When the mice were born, they carried the jellyfish gene in their own genes, and under fluorescent light all their major tissues and organs emitted a glow.

Many applications

Genetic engineering is used in medicine, agriculture, industry, and research. One of the most successful medical procedures has been the mass-production of the protein insulin, used for controlling diabetes. The human insulin gene is inserted in plasmids inside bacterial cells, and the bacteria are then cultivated. Although each bacterium produces only a tiny volume of insulin, billions of bacteria can produce a limitless supply. Genetic engineering has also been used for making antibiotics such as penicillin and vaccines for the control of a variety of diseases. In agriculture, crops have been genetically engineered to increase their resistance to pests, diseases, harsh environmental conditions, and herbicides.

87 Gaia Hypothesis

THE GAIA HYPOTHESIS WAS ARTICULATED BY THE BRITISH SCIENTIST JAMES LOVELOCK. It proposes that Earth's living and non-living parts form an interacting system that can be thought of as a single organism.

Planet Earth's capacity for self-regulation is being reduced by anthropogenic (man-made) damage to the environment— that is the view of the Gaia Hypothesis.

Lovelock proposed the Gaia Hypothesis in a 1979 book. The theory was named after the Greek earth goddess Gaia at the suggestion of Lovelock's friend, the *Lord of The Flies* author William Golding. It proposes that the biosphere has a regulatory effect on Earth's environment that helps to sustain it. In 1961 Lovelock was working at NASA, where he helped to develop precision instruments to analyze extraterrestrial atmospheres in anticipation of Moon and Mars landings. While Lovelock was pondering the prospects of life on Mars, he began to think anew about how Earth's biosphere might be operating. Lovelock's Gaia Hypothesis eloquently argued that planet Earth acts as a self-regulatory unit in which there are almost infinite interactions between living organisms and inanimate things such as the soil, air, and sea. Just as a person's body regulates its temperature, hydration, and blood sugar within the narrow range required to keep it alive, so Earth does the same. While he was working for NASA, Lovelock noted that while the proportion of gases in our planet's atmosphere is constantly changing— unlike what we have since learned about Mars and other planets—the fluctuations remain in a very narrow range that has not varied over much of Earth's history. Lovelock

DAISYWORLD

In 1983 Lovelock created a simulation of an imaginary planet called Daisyworld, which had only two lifeforms: Black daisies, which absorbed the Sun's heat and made the planet's surface hotter, and white daisies, which reflected heat back into space. The proportions of daisies oscillated with Daisyworld's temperature. Lots of black flowers warmed the planet, since they hang on to heat, thus making it possible for white daisies to thrive since they shed unneeded heat better. This increase in white flowers cooled the planet, making it easier for black daisies to thrive.

found similar stability in global temperatures and ocean salinity, but the mechanisms he proposed for how self-regulation operates are hard to verify. Nevertheless, many people found the idea that Earth was one complex system, like a living body, compelling.

Science or spiritualism?

Lovelock's theory wasn't well received by most of the scientific community, and many scientists suspected that Lovelock was proposing that Earth was a higher form of life, some kind of superorganism. Lovelock has been circumspect about this claim—his hypothesis suggests that the planet shares similar characteristics to an organism rather than actually being one—but the latter idea was the one that appealed to many who saw it as a scientific validation of Earth-inspired spiritualism. So the reaction from the "New Age" counterculture was much more positive. Many environmentalists also welcomed the theory. The supporters of the Gaia Hypothesis are not alone in arguing that modern civilization is disrupting the balance of Earth—and that we should act to ensure that the planet can regulate itself.

James Lovelock has argued that artificial climate change will produce increased temperatures, rising sea levels, and drought in parts of the world—rendering most of the planet uninhabitable during the 21st century.

88 Molecular Clock

EVOLUTION IS DRIVEN BY GENETIC MUTATIONS, WHICH ARE THE ULTIMATE SOURCE OF THE VARIATION ACTED UPON BY NATURAL SELECTION. The molecular clock uses the rate of mutation to gauge when organisms shared ancestors.

The molecular clock counts time by the accumulation of mutations in certain genes that are shared by a large range of organisms. Mutations arise by chance in individuals, and are mostly chemical changes in DNA, having no effect on the survival of the organism. However, they remain evidence of a relationship—organisms with this mutation are more closely related than those with a different gene. In the 1970s, biologist Motoo Kimura predicted that such mutations would accumulate at a certain rate. That rate could be used to calculate the time at which two related lineages shared common ancestry. Although the "clock" is not that accurate because the rate of mutation changes, geneticists can calibrate it for particular groups of organisms and use it to estimate the timing of major evolutionary events that are missing from the fossil record.

The molecular clock can be used to estimate when species diverged from one another by comparing how many mutations had accumulated in different genes.

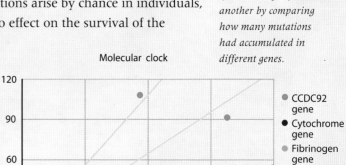

Molecular clock

- CCDC92 gene
- Cytochrome gene
- Fibrinogen gene

Number of mutations

Millions of years since divergence

89 Mass Extinction

MASS EXTINCTIONS ARE PERIODS WHEN ABNORMALLY LARGE NUMBERS of plant and animal species die out within a relatively short period.

The best-known mass extinction ended the "Age of the Dinosaurs" about 66 million years ago. It is known as the Cretaceous–Paleogene (K–Pg) extinction because it took place at the end of the Cretaceous period and the start of the Paleogene. In addition to the elimination of the previously dominant dinosaurs, an estimated three-quarters of all animal and plant species are believed to have become extinct, including the last of the flying pterosaurs as well as ammonites and many flowering plants.

The cause of the extinction had long been debated, but in 1980 a team of researchers led by Nobel Prize-winning physicist Luis Alvarez, his geologist son Walter, and chemists Frank Asaro and Helen Vaughn Michel made a dramatic discovery. Sedimentary layers at the K–Pg boundary all over the world contain a concentration of the element iridium hundreds of times greater than elsewhere. Iridium is very rare in the Earth's crust, so the Alvarezes drew the radical conclusion that it had come from an asteroid that had collided with the Earth. The collision, they argued, had thrown up enough dust to blot out sunlight and reduce plants' abilities to photosynthesize—creating an "impact winter." In turn, this reduced the food available to animals.

A decade later, geologists confirmed the existence of a huge crater straddling the coast of Mexico's Yucatan Peninsula. The size of the crater indicated a rock 10 km (6 miles) across had struck. Clearly this would have had devastating results, supporting the Alvarez theory. There is also reason to believe that a period of widespread volcanic activity which produced a vast sheet of basalt in southern India—known as the Deccan Traps—also contributed to the reduction of sunlight and to the ensuing K–Pg extinctions.

The impact that created the Chicxulub Crater is estimated to have created a shock one billion times the size of the Nagasaki and Hiroshima atomic bombs. The dust clouds created may have obscured the Sun for months.

Banded ironstones are common in Precambrian rocks dating to the time of the Great Oxygenation Event, about 2.4 billion years ago.

The Great Dying

Devastating though it was, the K–Pg extinction was not the biggest. About 250 million years ago, at the end of the Permian period, a time when most of Earth's life lived in the ocean, an estimated 96 percent of all marine species were killed off. In addition, 70 percent of land-living animals and plants died out. So much biodiversity was eliminated that it took millions of years for life to recover. Possible causes include an as-yet-unknown asteroid impact; the massive volcanic episode that released the basalt lavas of the Siberian Traps; a catastrophic release of methane from the seafloor; a shift in ocean currents—or maybe a combination of all these.

GREAT OXYGENATION EVENT

Oceanic cyanobacteria are possibly the first organisms to produce oxygen by photosynthesis. For millions of years the oxygen they produced was captured by dissolved iron in the oceans to form iron oxides. These can be seen in ironstone bands in ancient sediments. About 2.4 billion years ago, however, these oxygen "sinks" reached their capacity, and free oxygen was released into the atmosphere, with profound results. The oxygen was toxic for Earth's plentiful anaerobic organisms (mainly tiny organisms such as bacteria), so it was probably responsible for one of the greatest mass extinctions.

Other extinctions

The third-largest mass extinction took place at the end of the Ordovician period, between 455 and 430 million years ago. Believed to have been the result of the Earth's atmosphere cooling dramatically, glaciers would have advanced, and sea levels would have dropped. These environmental changes would have affected organisms on land, but especially in the oceans, where trilobites, brachiopods, and graptolites were hard-hit. Controversy surrounds the disappearance from the fossil record of many species in late Devonian times, 375–360 million years ago. This may have been the result of a series of extinctions rather than a single mass extinction. Life in shallow seas suffered catastrophically, and corals took many millions of years to recover. Marine lifeforms such as gastropods, brachiopods, bivalves, and cephalopods also succumbed in a mass extinction at the end of the Triassic period, about 200 million years ago. Many terrestrial vertebrates also died out, and the dinosaurs evolved to take their place.

Trilobites were abundant aquatic invertebrates, but were victims of three different mass extinctions, finally being wiped out at the end of the Permian period about 251 million years ago.

90 Polymerase Chain Reaction

THIS CHEAP, RELIABLE TECHNIQUE IS USED TO REPLICATE SEGMENTS OF DNA. It is used in genetic engineering, for the diagnosis and monitoring of hereditary diseases, and in DNA fingerprinting.

Polymerases are enzymes that copy DNA when cells divide. Invented by Kary Mullis in 1983, the polymerase chain reaction (PCR) method typically duplicates sequences of 100–10,000 pairs of DNA bases. Because part of the PCR process takes place at a high temperature, the DNA-polymerase is taken from thermophilic (resistant to heat) bacteria. First, a section of double-stranded DNA is heated to 96°C (205°F) to separate the two strands, then the temperature is reduced, and primers are placed to mark the start and end of the sequence to be duplicated. Finally, the DNA-polymerase replicates the missing bases in that sequence. The process is repeated many times over, and one billion copies can be produced in four hours.

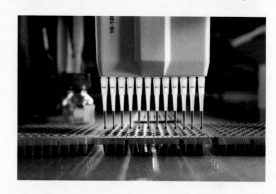

Preparing samples of DNA for copying in a PCR lab. The technique is used to diagnose diseases, identify viruses and bacteria, match criminals to crime scenes, and in many other ways.

91 Homeobox Genes

A HOMEOBOX, OR HOX, IS A SEQUENCE OF DNA, about 180 base pairs long, that is involved in the anatomical development of animals and plants.

Like all insects, the fruit fly Drosophila *has eight homeobox genes. The color coding indicates which gene acts on which body segment.*

In 1983, teams of researchers working independently at the University of Basel in Switzerland and Indiana University in the USA discovered that hox genes play a crucial role in the formation of body structures. The genes code for proteins that regulate other genes during early embryonic development. If there are mutations in the genes, body parts may be misplaced. For example, some mutant fruit flies have an extra pair of wings, and some have a pair of legs on their head in place of antennae. This happens when the gene instructs cells that would normally form antennae to form legs instead. Researchers in Australia have discovered the hox gene mutation that is related to intellectual disability in humans. Now, with the aim of finding ways to eliminate or correct problems, they are trying to understand how mutations alter development.

lab   pb   Dfd Scr Antp   Ubx   abd-A   abd-B

92 DNA Fingerprinting

DNA FINGERPRINTING IS A METHOD USED TO IDENTIFY AN INDIVIDUAL FROM A **SAMPLE OF GENETIC MATERIAL** by looking at unique patterns in their DNA. It has become an important forensic-science technique.

While British geneticist Alec Jeffreys was conducting research in his laboratory at Leicester University in 1984 he noticed that X-ray images of DNA samples from a family group (it was his technician's family) showed many similarities, but also some differences. He quickly realized a potential use for solving crimes. Although 99.9 percent of human DNA sequences are the same in every person, there are still enough differences that it is possible to distinguish between one individual and another, unless they are identical twins. If, for example, a sample of DNA taken from blood or other body tissues at a crime scene is compared with a sample taken from the saliva or blood of a suspect, it is possible to tell whether they are from the same person.

Jeffreys and his team went on to develop DNA profiling, which focuses on highly variable sections of DNA called minisatellites. By concentrating on just a few of these minisatellites that are unique to each individual, DNA profiling made the system both sensitive and easy to analyze on computer databases. Even shorter sections of DNA called microsatellites, or "short tandem repeats," are now used. In 1987 police were able to use the technique to convict a man for the murder of two young women. It is now the standard forensic DNA system used in paternity testing and criminal case work worldwide.

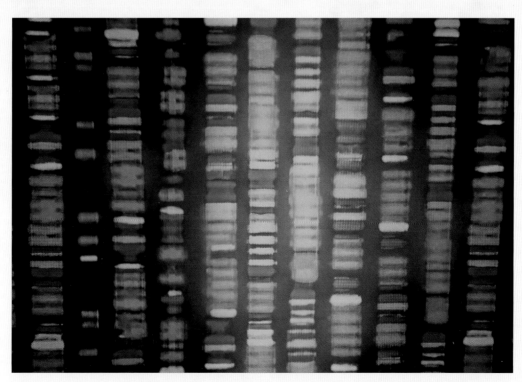

A genetic profile or fingerprint highlights sections of DNA that have repeated strings of the same base "letter." These "tandem repeats" are amplified and separated into a unique pattern —but a pattern that is also similar to that of a close relative.

93 Biodiversity

BIODIVERSITY REFERS TO THE TOTAL VARIETY OF ANIMALS, PLANTS, and other lifeforms in an area or across the whole of Earth's biosphere, including diversity within species, between species, and of ecosystems.

Although they cover less than two percent of Earth's surface, rainforests are home to about half of all known species of organisms.

The term biodiversity was coined by the American biologist Edward O. Wilson in the 1980s. The concept soon became widely understood beyond the scientific community, reflecting fears about the future of Earth's diversity. At the simplest level, biodiversity refers to the total number of species and subspecies of organisms. (Subspecies are genetically distinct populations of an organism that have not become genetically different enough to be considered full species.) About 1.9 million separate lifeforms on our planet have so far been described, including more than 1 million insects, about 310,000 plants, and 65,000 vertebrates. Mammals make up a relatively tiny fraction—fewer than 5,500 species. Many organisms are yet to be discovered, and although estimates of their number vary, a figure of 8 million is generally accepted. Areas of greatest biodiversity include tropical rainforests, tropical coral reefs, and the Cape floral region of South Africa. Biodiversity has always been affected by a wide range of non-human factors, including natural climate variation, sea level change, and even asteroid impacts, all of which have been responsible for mass extinctions. Currently, there is evidence that human activity, including habitat destruction and fragmentation, overexploitation for food, climate change, pollution, and the impact of invasive species, is responsible for the largest spike in extinctions since the time of the dinosaurs.

Coral reefs have an extraordinary biodiversity, but that variety is threatened by increasing water temperatures. More than a quarter of oceanic species live around reefs.

94 Domains

LINNAEUS DIVIDED LIFE INTO TWO KINGDOMS—PLANTS AND ANIMALS. Today the highest classification rank of organisms is the domain. Most scientists accept that there are three domains, though some propose five.

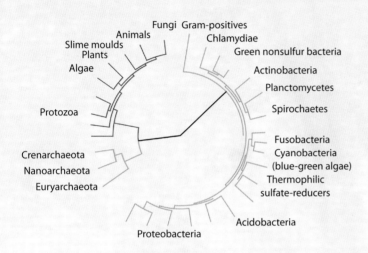

Carl Woese's three-domain classification, with Archaea shown in green, Bacteria in blue, and Eukarya in red. Animals make up just one of dozens of lineages.

Most biologists now recognize five kingdoms of life: Monera (bacteria), Protista, Fungi, Plantae, and Animalia; but domains were added above kingdoms by American microbiologist Carl Woese. In 1977 he found that prokaryotes (single-celled organisms without organelles) exist in two distinct groups. Those living at high temperatures or producing methane have a different genetic make-up from bacteria and eukaryotic life. Woese had discovered an entirely new kind of life, the Archaea.

Archaea, Bacteria, and Eukarya

In 1990 Woese proposed that all life should be divided into three domains, which we now know as Archaea, Bacteria, and Eukarya. Each is marked out by distinct ribosomal ribonucleic acid (rRNA) within its cells. All members of the Archaea are tiny, single-celled organisms in which the cell doesn't have a nucleus. Examples of archaeans include halophiles, which thrive in very salty water, and hyperthermophiles, which tolerate high temperatures. The cells of these microorganisms are adapted to deal with the harsh conditions. Bacteria also lack a cell nucleus, but the structure of their rRNA differs from that of the Archaea. The Bacteria domain is very varied and includes photosynthetic cyanobacteria and rock-eating lithotrophs. The Eukarya is also a diverse domain, which includes plants, animals, fungi, and single-celled protists. Viruses, which do not have a cell, were omitted from Woese's classification. Not everyone agrees with the three-domain division. In 2012 the Swedish microbiologist Stefan Luketa proposed a five-domain alternative, adding prions (infectious proteins) and viruses to Woese's categories.

EXTREMOPHILES

In the 1970s, scientists began to find organisms living in conditions previously considered inhospitable. Most are microorganisms, especially bacteria, but there are also worms, insects, and crustaceans. These "extremophiles" include acidophiles, which thrive in acidic environments with a pH of 3 or lower—strong enough to burn the skin. Hydrothermophiles live in water temperatures above 80°C (176°F) close to hydrothermal vents on the ocean floor. Others live deep underground, beneath glacier ice, in extremely alkaline conditions, or around boiling geysers.

95 Cloning

SEVERAL PROCESSES USED TO PRODUCE GENETICALLY IDENTICAL COPIES of a biological entity are known as cloning. There are three methods. In 1996, a sheep called Dolly was the first cloned mammal to be produced.

Clones are common in nature. For example, plants and single-celled organisms that reproduce asexually produce offspring with identical genetic make-ups. And identical twins are clones of each other. However, the term cloning generally refers to techniques to make artificial copies of genetic material and organisms. Gene cloning is a method of producing copies of DNA for use in medicine and research. Therapeutic cloning is proving to be very useful in medical research because it produces embryonic stem cells. Stem cells build, maintain, and repair the body, and since this is what they do naturally, they could be manipulated to repair damaged or diseased organs. However, stem cells transferred from one person to another are seen as "foreign" by the host body and usually trigger an immune response. Scientists are looking at cloning as a way to create stem cells that are genetically identical to the individual who could benefit. In future they could be used to repair tissues, or even replace whole organs.

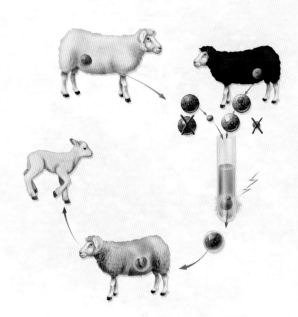

This diagram explains the main stages of reproductive cloning that made Dolly. A cell was taken from the udder of Dolly's biological mother. The nucleus was removed from the ovum of another female sheep. Then the cell nucleus from Dolly's mother was inserted into the ovum, which was shocked into action to make it divide. The ovum—now an embryo—was placed in a third sheep, its surrogate mother. Dolly was then born in the normal way at the Roslin Institute, Edinburgh, Scotland.

Dolly the sheep

Reproductive cloning uses similar methods as therapeutic cloning, but produces copies of whole animals. The most famous cloned animal is Dolly the sheep, who was born in July 1996 and was the first clone created from an adult mammal.

Geneticists Ian Wilmut and Keith Campbell directed the project. Despite their success, reproductive cloning is very difficult. Dolly was the only survivor from 227 attempts. Other cloned animals have included fish, pigs, cats, rats, and a dromedary camel. A gaur, an endangered type of wild cattle, has been cloned, showing the potential for boosting populations of rare animals and even recreating extinct ones. Cells are being collected from endangered animals in the hope that if their species becomes extinct, the cells can be cloned.

After her healthy birth, Dolly the sheep gave birth in turn to six lambs. She only lived for six years, dying young from a lung disease. This was probably caused by spending so much time indoors—she was very valuable and could not be left out in a field. However, it is also suggested that Dolly's chromosomes were actually older than she was—they had not been renewed as they passed from the mother—and she may have had a shorter life as a result.

96 Human Genome

THE COMPLETE NUCLEIC ACID SEQUENCE FOR HUMANS, encoded as DNA, is described as the human genome. It is made up of the genetic material in the 23 chromosome pairs and in the cell's mitochondria.

The Human Genome Project plotted an almost complete sequence of all 3 billion base pairs in human DNA. The project started in 1990, collecting DNA from a number of anonymous donors, with about 70 percent coming from one man in Buffalo, New York. The first draft was published in 2001. While there are significant differences between the genomes of individuals (apart from identical twins), this amounts to about 0.1 percent of the total genome (but that is still millions of differences). The difference between humans and our closest relatives, chimpanzees, amounts to around 4 percent. Humans have 19,000–20,000 protein-coding genes, but the biological function of their protein and RNA products is far from understood. By identifying the function of genes, the expectation is that it will lead to advances in the treatment of disease, as well as giving new insights into the history of human evolution.

The DNA in human mitochondria (mtDNA) could offer even more clues about our ancestors. Since it is not subject to the same copy-checks as the DNA inside the cell nucleus, its mutation rate is much higher. Analysis of mtDNA has allowed ancient paths of human migration to be traced. In 2016 a study of the genome suggested that all non-Africans can be traced to a single population that left Africa 60,000 years ago.

Part of the genome map of chromosome 16, which has about 90 million base pairs, representing about 3 percent of the total DNA in human cells. Chromosome 16 contains about 2,000 genes.

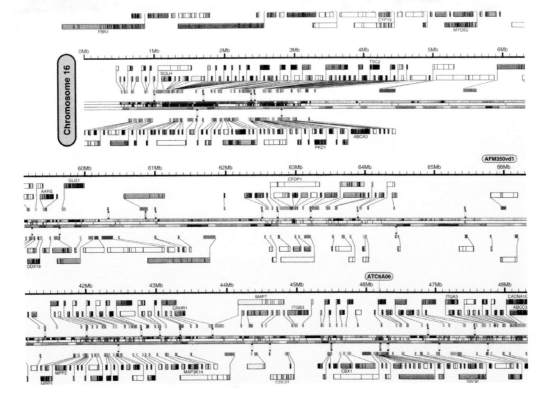

97 Epigenetics

IF AN ORGANISM'S GENETIC CODE IS THOUGHT OF AS A SCRIPT OUTLINING WHICH GENES SHOULD BE SWITCHED ON OR OFF, epigenetics is the study of how this gene expression is regulated by environmental factors.

The genetics revolution changed biology, but it also sparked popular confusion. Could we escape our genes? Or was everything about us pre-programmed? Then epigenetics arrived and made it even more confusing. It is a central axiom (or rule) of genetics that inherited genetic material—the DNA that codes our genes—remains unchanged by the body. Whatever you inherited from your parents, you will pass on to your children. That fact underlies everything we know about genes and is the central tenet of some views of evolution, population genetics, and developmental biology. (Of course mutations, or mistakes, in genes do occur but this is by accident during copying or by an outside attack altering the chemistry of DNA, not through the action of a process started by the body.)

So in other words, all this means that, according to the field of genetics, you only inherit genetic characteristics from your parents. You do not inherit any characteristics they may have acquired during their lives—things like big muscles from weightlifting, or sore feet from badly fitted shoes.

There are between 2 and 3 meters (roughly 6–9 ft) of DNA in every human cell. Those long, flimsy chemicals need careful storage, a job done by histone proteins, which coil the DNA into ultra-compact units. The structure of the histones and other supporting chemicals forms the epigenome, which is inherited along with the genome.

Epigenome

But a new field called epigenetics suggests another element is at work during inheritance. While the genetic axiom still holds firm, the way the body develops and makes use of its genes may be controlled by factors that are acquired in life—and may be passed down from a parent (at least from the mother). Central to this idea is the epigenome. This refers to the army of chemicals, such as histones, that surround, support, and manage DNA inside a cell. (The DNA is called the genome.) Experience in life—poor diet, plenty of exercise, even diseases—may alter the epigenome in significant ways. This has the effect of locking some genes away from use by a cell and opening up others that would otherwise be unused. While the genome is passed from parent to offspring unchanged, the epigenome that travels with it might be totally different to the one the parent inherited themselves.

DUTCH HUNGER WINTER

In the winter of 1944–5, the German army stopped food reaching the Netherlands, creating famine conditions. The *Hongerwinter* (Hunger Winter) affected 4 million people, and in the years after the war, many of them —and their offspring—were monitored for long-term health effects. The results offered some of the first evidence of epigenetic inheritance.

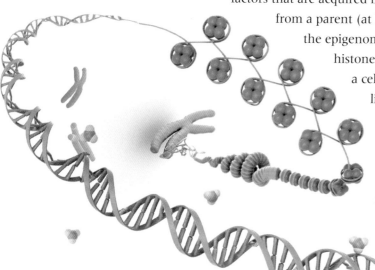

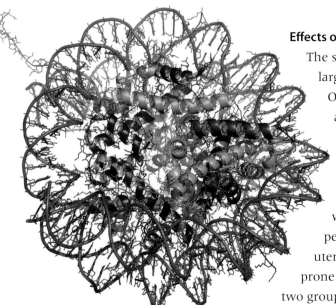

Effects of famine

The strongest evidence for epigenetic effects comes from large-scale studies of families over several generations. One of the biggest followed the Dutch Hunger Winter, a period of famine in the Netherlands at the end of World War II. Babies that were born during or just after the famine form the basis of the study, along with their parents, children, and soon grandchildren, too. It appears that babies born during the famine were generally small and thin in adult life (but perfectly healthy). Those born just after—they were in utero during the famine—grew up to be overweight and prone to mental illnesses. Intriguingly, the children of the two groups share the same characteristics as their mothers. It appears that the effects of malnutrition were felt by the mother, the developing baby, and also the cells inside female babies that would one day become the eggs used to make the next generation. All three generations, mother, child, and grandchild, were impacted by the famine. In a few years, data will be available about the fourth generation. Will the epigenetic effect have been passed on that far?

The histone proteins that support and organize the long strands of DNA in a cell were once thought to be purely structural supports. However, epigenetics suggests that they have a role to play in inheritance, at least in the short term.

98 Cyborgs

THE SCIENCE FICTION CYBORG, A PART-HUMAN, PART-MACHINE BEING, has been popular since the 19th century. Now cyborgs are a reality, using appliances to replace or improve body functions.

This cyborg moth carries a tiny computer that helps researchers detect the activity of its antennae as they pick up scents—which the moth flies toward. This may lead to a system where the moth can be controlled remotely as a tiny cyborg drone.

Body enhancements, ranging from prosthetic limbs to hearing aids and contact lenses, have been used in medical science for many years. The advent of microelectronics has made the possibilities much greater. After electrician Jesse Sullivan had both arms amputated in an accident in 2001, the Rehabilitation Center of Chicago turned him into a cyborg by fitting him with a fully robotic arm, which he can operate by thought. Nerves from his shoulder, which once controlled his arm muscles, were grafted on to his pectoral muscles. When the nerves receive thought-generated impulses they cause muscle contractions that are detected by electrodes. These relay signals to his arm's computer, which causes motors to move his elbow and hand. Other cyborg appliances include artificial kidneys and stomachs, retinal and cochlear implants to provide replacement vision and hearing, 3D joint systems, and artificial skin.

99 Synthetic Biology

SYNTHETIC BIOLOGY COMBINES BIOLOGY WITH CHEMISTRY, NANO-ENGINEERING, AND COMPUTER SCIENCE. It can replicate and accelerate natural processes, and even build artificial lifeforms.

Synthetic biology takes natural processes and tries to improve them. Photosynthesis, for example, makes useful chemical energy out of sunlight. However, it is very inefficient—only about one percent of the Sun's energy is converted into sugars. Synthetic biology is looking at ways of increasing that efficiency, so artificial algae or bacteria, equipped with enzymes from re-engineered genes, could become a source of organic material produced by photosynthesis. The material could be used to make fuels or as feedstock for the chemical industry. This is one way we could reduce our reliance on petroleum, without needing to devote large amounts of land to grow plants for biofuels.

Synthetic biologists have succeeded in making artificial cells assembled from components taken from other cells. These cells are able to divide and grow just like natural ones. In 2012 bioengineers also developed XNA, a synthetic version of DNA that works in the same way but is much more robust. (DNA, though miraculous, can be somewhat fragile.) Artificial XNA—that stands for xeno nucleic acid—has a stronger chemical framework while leaving the code-carrying elements untouched. Potentially, XNA organisms could go where DNA-based life cannot.

CHEMICAL EVOLUTION

Artificial nucleic acids, like XNA, are based on chemicals that may have existed at the dawn of life. The simplest lifeform is a strand of RNA. This can autocatalyze, meaning it can use itself as a template to make a copy, or mirror image, of itself—a primitive form of reproduction. How RNA and DNA evolved is one of the mysteries of biology. It is assumed that they are the last surviving chemicals among a range of similar autocatalyzing polymers that developed 3.8 billion years ago. These chemicals were subject to natural selection in the period of chemical evolution. Biochemists have synthesized many such chemicals to investigate how they might have behaved. XNAs are based on versions of these test chemicals.

Synthetic biology might one day refit cells with XNA to make an entirely new domain of life. In addition, XNA could be used in gene therapies in the body.

100 Astrobiology

DOES LIFE EXIST BEYOND EARTH? THE SCIENCE OF ASTROBIOLOGY SEEKS TO ANSWER THAT QUESTION. No aliens have been found, but the number of places known to have the potential to support life is growing steadily.

Jupiter's moon Europa is considered the most likely location for terrestrial life in the solar system. The ocean under its icy surface might host mineral-eating microorganisms similar to the ones that live around hydrothermal vents. These bugs could form the foundations of an alien food chain.

When NASA was established in 1958 one of its aims was to learn how to look for the presence of life beyond Earth. Since then, although extraterrestrial evidence has proved hard to come by, a great deal has been learned to assist the search. For example, in 1977 previously unknown communities of microorganisms were discovered thriving alongside boiling hydrothermal vents on the ocean floor, conditions previously considered uninhabitable. Discoveries from other extreme environments suggested that life could exist beyond Earth's biosphere—such as in the frozen-over ocean of Jupiter's moon Europa. Researchers have also found all the chemicals needed for life in space. Amino acids, key building blocks for life, were found in samples of the comet Wild 2 after NASA's Stardust spacecraft passed through the comet in 2004. This indicates that the ingredients for life are on planets, moons, and asteroids everywhere.

The actual discovery of extraterrestrial life will depend on space telescopes and robotic missions. Guided by the idea that where there's water, there may be life, scientists have found potential within our own solar system. Jupiter's moon Europa has an ocean beneath its icy crust, and water has also been detected spurting out of Saturn's moon Enceladus.

Astrobiologists are also looking at planets orbiting distant stars (exoplanets). Some of them are rocky and orbit at a distance from their sun where water could be liquid. To find out, scientists will need to study the atmosphere of these planets, and look for the tell-tale chemicals produced by life. In 2018, NASA will launch the James Webb Space Telescope, the first instrument able to detect chemicals in alien atmospheres.

SPACE BUGS

Tardigrades, or water bears, are micro-animals that are extremophile—they can survive almost anywhere. They were the first animals known to survive in space when, in September 2007, dehydrated tardigrades were taken into orbit. Despite being exposed to the vacuum of space for 10 days, more than two-thirds of them revived within half an hour after being rehydrated back on Earth.

Biology: **The basics**

BIOLOGY IS A BIG SUBJECT. IF IT'S ALIVE—OR WAS ALIVE ONCE—A BIOLOGIST will investigate it. With such a broad sweep of topics on offer, what basic tenets can we find in biology? The only place to start is with life itself.

What is Life?

Movement Every living thing, from a blue whale to a blueberry bush and bacteria does seven basic things. The first is they all move. It's easy to see when a leopard is on the move, or another animal, but how do plants move? Plant motion is generally much slower, admittedly, but, as they grow, they are moving—too slowly to see—toward a source of light and away from the pull of gravity.

Excretion Life maintains its internal conditions through a phenomenon called homeostasis. A universal feature of this is getting rid of waste, a process called excretion. The white stuff that  birds excrete is actually a solid urine (their feces are darker). Meanwhile, plant excreta is oxygen—and we breathe that.

Respiration

All life needs a supply of usable energy, and this is produced by a process called respiration. Most lifeforms use oxygen to release packets of energy from fuels, which are usually sugars such as glucose. Respiration occurs in every cell; large cells use little power plants called mitochondria to do it. However, it is possible to live without oxygen. For example, gut bacteria are killed by air and get their energy by fermenting fuels.

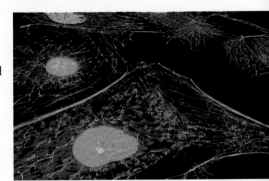

Reproduction

All life has some way of making a copy of itself. Creating offspring is a way for an individual's genetic material to proliferate and persist into the future. Reproduction can be as simple as splitting in half, which is an example of asexual reproduction. Sexual reproduction involves a mixing of genetic material from two parents, which requiries more investment of time and energy to be successful.

Sensitivity

All organisms are sensitive to their surroundings, and most crucially, respond to changes. In animals, this is manifested as sense organs, such as the eyes and ears—although there are also senses for motion, chemicals, and even electricity. Plants are also able to sense light, and grow toward it, while bacteria will enter a dormant state when conditions turn against them, forming a tough, protective cyst around their cell.

Nutrition

Life requires a source of nutrition, which provides the fuels that drive metabolism. Plants get nutrition from photosynthesis, while animals and other heterotrophs will consume nutrients (including other organisms) as food. Plant nutrition also entails collecting minerals that cannot be synthesized inside the plant. Commonly, this is done by roots harvesting minerals from the soil.

Growth

Another feature of life is growth. This is obvious in multicellular organisms, which develop through various stages of a life cycle to reach a mature state, when they are able to reproduce. In single-celled organisms, it can be difficult to differentiate growth from reproduction. Binary fission, or when the cell splits in half, is most like growth, while reproduction might involve a sexual component at some stage.

Fields of Biology

Biology is a broad science—perhaps the broadest of them all. Scientists who fall under the umbrella term of "biologist" may be studying the chemical interactions of proteins in a cell, the flow of nutrients through an entire ocean or jungle, the vocal expressions of a prairie dog, or the possible inhabitants of a moon of Saturn. Let's look at just a few specialist fields.

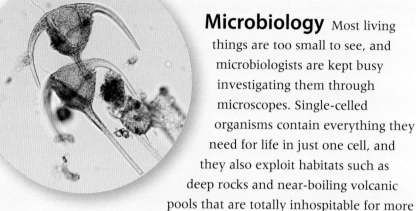

Microbiology Most living things are too small to see, and microbiologists are kept busy investigating them through microscopes. Single-celled organisms contain everything they need for life in just one cell, and they also exploit habitats such as deep rocks and near-boiling volcanic pools that are totally inhospitable for more familiar forms of life. Nevertheless, these kinds of habitats are thought to be where life began.

Taxonomy This is the field of biology that seeks to figure out how all life on Earth is related. It's a big job, considering that about 80 percent of species have yet to be described and given a name. The field of taxonomy began by linking organisms that looked alike and lived in similar way. Modern taxonomists also take into account similarities in how organisms develop from embryos, and compare genes to establish relationships.

Ethology An ethologist is also known as an animal behaviorist. Ethology involves observing animals in their natural habitats over a long period to see what they get up to. However, the science also demands stringent statistical analysis of behaviors in controlled conditions. This is the only way a natural behavior can be proven to perform a particular function.

Ecology The study of natural habitats and the organisms that live in them is called ecology. Ecologists want to know the relationships that exist in wildlife communities, often so they can protect them from outside interference by human activity. Ecology involves sampling habitats to create a map of species by location and population density.

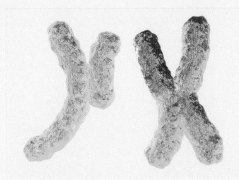

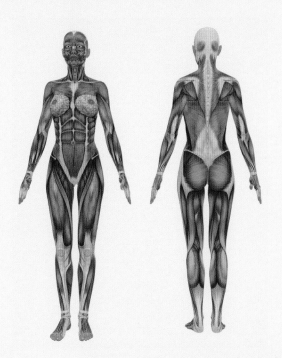

Genetics

The study of heredity, genetics is focused on genes—how they are formed, how they are transmitted by inheritance, and, most importantly, how they are translated into a measurable trait that can be seen in an organism. Geneticists have developed advanced technological tools to read genes on strands of DNA. Genetic engineers are able to alter DNA to create lifeforms not seen in nature.

Physiology

Often associated with the human body, physiology is the study of a body's mechanisms in order to understand their normal functions. This involves investigating the material properties of living tissues, such as bone and muscle, but also looks at the physical behaviors of all living tissues. Physiologists also study periodic changes in a body, caused by changes in season, temperature, and other exterior influences.

Cell biology

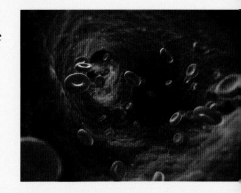

The cell is the unit of life—all organisms are made of at least one cell, and all new cells arise from an older one. Cell biology focuses on what is going on inside a cell and on how different cell types function to form tissues and organs within a body. Cell biologists, sometimes called cytologists, contribute to genetics, physiology, and biochemistry.

Biochemistry

Biology is driven by chemistry. A single cell plays host to several thousand individual chemical reactions. Together, these reactions make up metabolism, and biochemists look to unravel its complexities. Biochemists research a vast range of compounds, such as lipids, carbohydrates, and proteins, modeling their structure to understand how they function.

Evolutionary biology

This field of biology seeks to understand where the great diversity of life seen in Earth's present and past came from—and even suggests how it might change in future. The great majority of biologists subscribe to the theory of evolution by natural selection, but that well-established theory does not provide a full answer to the way organisms change over time.

Cell division

All life is made of cells, and all new cells are made from older cells that have divided. There are two kinds of cell division seen in nature, both of which follow a complex set of stages. Mitosis drives growth, while meiosis allows for genetic material to be mixed during sexual reproduction.

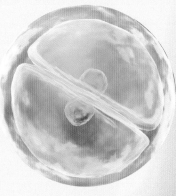

Cell division requires that the nucleus divides, too.

Meiosis This process is two cell divisions resulting in four daughters from one parent. The daughters contain half the amount of DNA as the parent. Sex cells, such as sperm and eggs, are made this way.

Interphase
Cell translates genes into proteins, performing normal metabolic processes.

Prophase
Chromosomes duplicate into two connected chromatids. A spindle begins to form.

Mitosis This cell division produces two daughter cells from one parent. The daughters are identical to each other—and to the parent. A cell spends most of its time in interphase. This is when the cell is busy metabolizing. Mitosis starts with prophase, the first of five steps.

Parent cell

Prophase
Chromosomes duplicate into two connected chromatids.

Centrioles (organelles involved in dividing the cell) begin to move to opposite ends of cell.

Prometaphase
The chromosomes thicken and the nuclear membrane begins to break down.

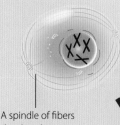

A spindle of fibers develops between the centrioles.

Metaphase
The chromosomes line up along the middle of the cell.

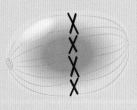

Daughter cell

Telophase
New membranes form around the chromosomes. A cell membrane forms across the middle of the cell, which then splits in two in a process called cytokinesis.

Anaphase
The spindle pulls the chromatids apart, making them into separate chromosomes, which move to opposite ends of the cell.

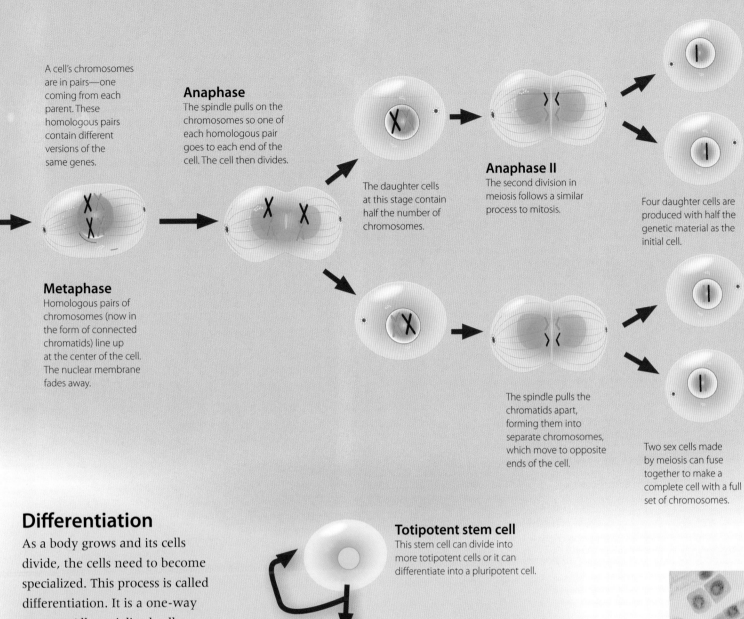

A cell's chromosomes are in pairs—one coming from each parent. These homologous pairs contain different versions of the same genes.

Anaphase
The spindle pulls on the chromosomes so one of each homologous pair goes to each end of the cell. The cell then divides.

Metaphase
Homologous pairs of chromosomes (now in the form of connected chromatids) line up at the center of the cell. The nuclear membrane fades away.

The daughter cells at this stage contain half the number of chromosomes.

Anaphase II
The second division in meiosis follows a similar process to mitosis.

Four daughter cells are produced with half the genetic material as the initial cell.

The spindle pulls the chromatids apart, forming them into separate chromosomes, which move to opposite ends of the cell.

Two sex cells made by meiosis can fuse together to make a complete cell with a full set of chromosomes.

Differentiation

As a body grows and its cells divide, the cells need to become specialized. This process is called differentiation. It is a one-way process. All specialized cells develop from dividing stem cells. Once specialized, they cannot be transformed back into stem cells or divide into another type of cell.

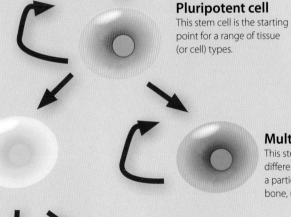

Totipotent stem cell
This stem cell can divide into more totipotent cells or it can differentiate into a pluripotent cell.

Pluripotent cell
This stem cell is the starting point for a range of tissue (or cell) types.

Multipotent cell
This stem cell is able to differentiate into the cells in a particular tissue, such as bone, nerves, or skin.

Blood stem cell
This stem cell is multipotent. It can produce the different cells seen in the blood system.

Red blood cells
Used for transporting oxygen.

Platelets
Involved in clotting.

White Blood cells
Involved in immunity.

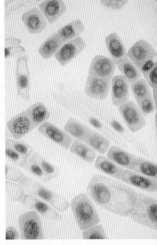

All eukaryotic cells (those with a nucleus and complex internal structure), including these plant cells, divide in the same way.

Biodiversity and biomass

There are two ways to measure the natural world. Biodiversity is the number of distinct species. The number of species on Earth far outstrips our ability to count them: It is thought that there are about 8 million, but this number could be a significant underestimate. Some authorities suggest that prokaryotic species alone might number more than 100 million. Another way of quantifying nature is biomass, which is the total weight of living material. The biomass of multicellular organisms is about 500 billion tonnes. The weight of prokaryotes is less well understood, with biomass estimates ranging from just 4 billion tonnes to as much as 300 billion tonnes.

Proportions of species

This chart shows the proportions of different kinds of multicellular species. The stand-out figures are that insects make up more than half of the total, and only one in a hundred species is a vertebrate.

Endangered species

Human activities have meant that thousands of animals are in danger of extinction, some critically so. Much of the attention in this regard is focused on large mammals. However, the best way to protect them is to protect their habitat, which also benefits less familiar species.

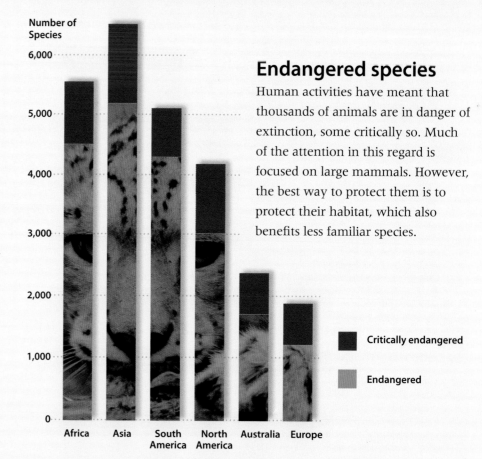

Number of Species

■ Critically endangered

▨ Endangered

Africa · Asia · South America · North America · Australia · Europe

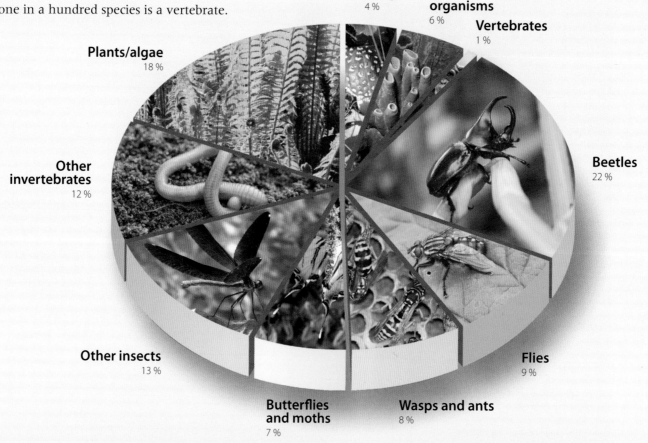

Fungi 4%

Other organisms 6%

Vertebrates 1%

Plants/algae 18%

Beetles 22%

Other invertebrates 12%

Other insects 13%

Flies 9%

Butterflies and moths 7%

Wasps and ants 8%

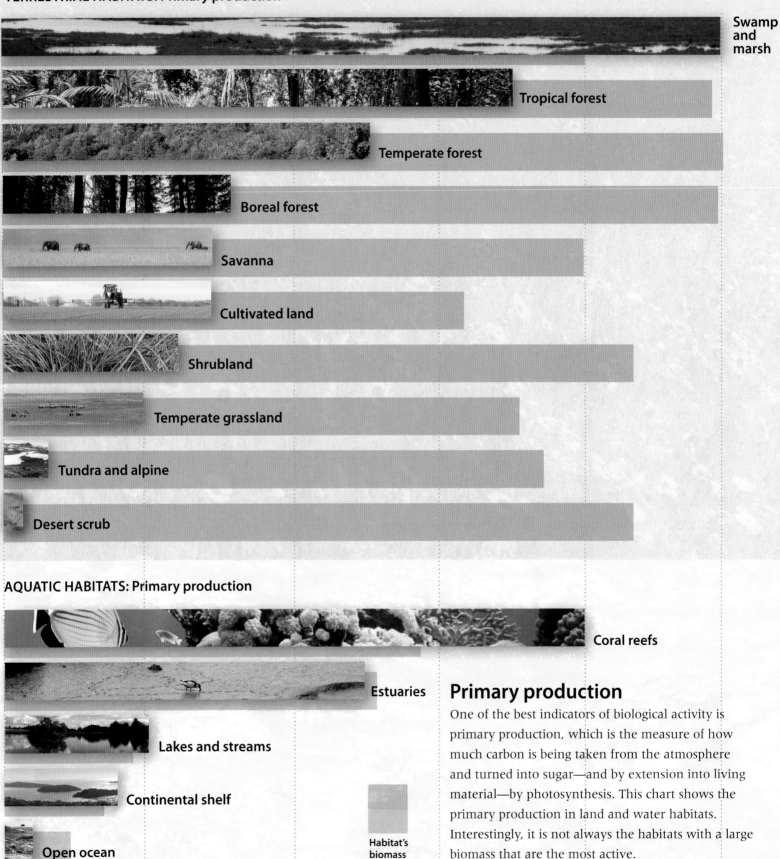

TERRESTRIAL HABITATS: Primary production

Swamp and marsh

Tropical forest

Temperate forest

Boreal forest

Savanna

Cultivated land

Shrubland

Temperate grassland

Tundra and alpine

Desert scrub

AQUATIC HABITATS: Primary production

Coral reefs

Estuaries

Lakes and streams

Continental shelf

Open ocean

Habitat's biomass

Primary production

One of the best indicators of biological activity is primary production, which is the measure of how much carbon is being taken from the atmosphere and turned into sugar—and by extension into living material—by photosynthesis. This chart shows the primary production in land and water habitats. Interestingly, it is not always the habitats with a large biomass that are the most active.

IMPONDERABLES

BIOLOGY HAS COME A LONG WAY SINCE PEOPLE THOUGHT A DOLPHIN WAS A FISH, OR THAT BABIES GREW FROM TINY PEOPLE INSIDE A SPERM. We now know about cells, inheritance, biochemistry, and evolution, but there are many problems that still need an answer. Let's ask a few of them.

Where did life start?

The fossil record tells us that the first life—bacteria-like creatures—appeared about 3.4 billion years ago. These ancient beasties had cells, and presumably DNA or RNA, with genetic material driving their survival and reproduction. This complexity did not arise from nothing, but must have been produced by an earlier process, where life did not use cells—in which case can we call it life? Perhaps. Nucleic acids, or at least precursors to them, arise through natural chemical processes taking place in an energetic medium—they were cooked up in a primordial soup of chemicals. However, there is a gap between these simple chemicals and the complex molecular chains that underwrite cellular life. The early Earth was indeed energetic and highly changeable, with volcanic eruptions and meteor strikes being much more common than today. It sounds strange, but hot springs, sizzling rocks, or spouting volcanic vents were among the most stable places around—extreme, but stable. Could it be that the long process of growing RNA and then DNA took root here? After all, these are the places where we find the most primitive organisms today.

Did life start down there?

What is aging?

Life spans vary enormously. Bristlecone pines live for more than 4,000 years, Atlantic quahogs (a type of big clam) have been clocked at 500 plus. Things don't die just because they get old, but because the body ages. It appears that aging is not simply the accrual of injuries and other insults over the years, but an active process. Chromosomes have structures called telomeres, which act as counters that limit the number of times a chromosome can be copied. This limiting factor is part of growth, as old cells die to make way for the new. No one is quite sure, but it may also rule over the process of aging, which rids the gene pool of well-used, well-copied DNA once the organism has passed through its reproductive years. If that's the case, then aging and death may not be an inevitable consequence of life.

Sea turtles outlive humans; most make it past 80.

Why do animals sleep?

Although it might not be quite like ours, all organisms go through a cycle of rest and activity, even bacteria. As far as we can tell, all animals sleep—certainly all vertebrates, including fish, frogs, and reptiles. Our experience of sleep suggests that we do it because our body needs a rest. It is a period of healing, certainly, but why do we have to become unconscious to do it? It may be a vestige of our evolution—did a distant ancestor sleep for some reason that no longer applies, but we still do it anyway? If that were the case, why has no animal evolved to not sleep? Just imagine the advantages. Sleeping uses 95 percent of the energy required for restful waking, so we are not saving much energy. Sleep researchers have shown that the brain remains active as we sleep. Perhaps it switches off consciousness to give it time to process memories from the day? Another idea is sleep is a time of physically flushing waste from the brain. There are a lot of options. Which one do you like? Why not sleep on it.

Sleeping might actually be important work. Let's get to it.

IMPONDERABLES

Was the Ediacara biota a failed form of life?

Think of an animal, any animal—or plant for that matter. Whatever you choose, its lineage can be traced back to the Cambrian period about 500 million years ago, when an incredible diversification of life took place in the shallow oceans. However, earlier fossils dating an extra 100 million years ago that were dug up in the Ediacara Hills of Australia show complex life existed well before this. However, this set of organisms, the Ediacara biota, has baffled fossil experts. It consists largely of lobed organisms with appendages. Are they fronds and stalks of some ancient sea plant, or the bodies of free-swimming jelly animals? Perhaps they were blobs of bacteria? The bigger question is, were they the ancestors of the organisms that crowded the Cambrian seas, or do they represent a first go at complex life, one that was doomed to failure?

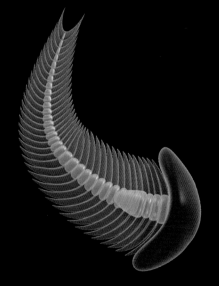

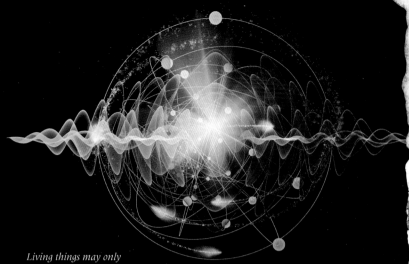

Living things may only work due to the wacky world of quantum physics.

Does life use quantum physics?

Quantum physics tells us about the nature of subatomic particles. They are packets of mass and energy that can behave like a wave and like a point of matter, and most crucially, they can be both at the same time. It is impossible to pin down an electron or other subatomic particle. Its many characteristics are expressed in terms of probabilities—the chances that it is moving this way or that, spinning one way or another, or located here or there. This idea is called superposition, which means that electrons and their ilk can be in more than one place at the same time and behave in more than one way at a time. All very interesting, if not a little baffling, but can biology learn anything from this way of describing nature? One field of biochemistry that has been set alight by quantum contributions is photochemical reactions. Our eyes see because the light photons ping into a chemical in the retinal cells, sending out a shower of electrons. Similarly, photosynthesis relies on this same kind of phenomenon to transfer energy from light to the cell. Enzymes use only a tiny amount of energy in their work, so perhaps they use quantum tunneling, which allows electrons to break through energy barriers. It may be that life relies on these quantum effects, without which the whole endeavour would fail.

What is the protein-folding problem?

Proteins are complex polymers of amino acids, all chained together in a specific order. The precise order is very important (and is coded for in genes) because once constructed, the protein will fold up into a unique shape. That shape is crucial for the protein to do its job in metabolism. That much we know, but predicting the shape of a protein from the order of its amino acids is something that we are still struggling to understand. Proteins fold by themselves or with the help of ions and other chemicals. They form a complex of weak bonds between different parts of the molecule, and the strength of each bond depends on the nature of other bonds around them. As such, even with computer modeling, predictions of protein folding are currently inexact—and if one thing needs to be exact in biology, it is the shape of proteins.

If we can crack the protein problem then bioengineers will be able to build new proteins for use in industrial chemistry.

What is intelligence?

Humans are undoubtedly intelligent. Just by reading that sentence you have proved it. Our primate brains have large frontal lobes for predicting events and making plans. However, is this scheming and dreaming form of intelligence—based on years of learning and experience—the only type? Octopuses never learn from their parents. They die without passing on knowledge to their offspring, but octopuses have remarkable problem-solving skills. Do they use the same modes of thought to do it as us, or do they have one of many other ways of understanding their situations?

Do animals have culture?

Culture is learned behavior and it is a common aspect of social animals. For example, orca pods hunt different prey—whales, fish, seals, etc—according to the pod's culture, while birds and even cattle have been shown to vocalize using accents or dialects associated with where they live. The question for biologists is how do these cultures emerge and spread? Distinct cultures are generally associated with populations that are isolated from each other. Some behaviors will be due to evolutionary adaptation, while others are from individuals innovating new ways of survival. Once in place, is it possible that cultural behavior can spread by natural selection?

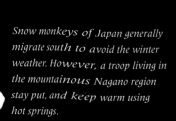

Snow monkeys of Japan generally migrate south to avoid the winter weather. However, a troop living in the mountainous Nagano region stay put, and keep warm using hot springs.

IMPONDERABLES

The suit protects the astronaut. But for how long?

Can humans live in space?

What is the near-death experience?

People who have been gravely ill, but survived, sometimes (although rarely) report a near-death experience. These experiences have common features such as seeing the body from above, meeting dead relatives, and traveling to another realm. Brain scans of dying rats reveal what is called the "death spike," which is a surge of activity in the temporoparietal junction brought on by a rise of carbon dioxide in the blood. This region of the brain is associated with the sense of self, and it is conjectured that the dying brain is trying to make sense of its situation, by combining any sensory information with memories from the past. As the sensory data begins to fall away, the self becomes disconnected from the body. What happens then is one question that cannot be answered.

Space exploration began as a race to take control of the orbits around Earth, and to telegraph the power and capability of a nation's rocket (and missile) technology. But it was also about exploration and science. The longest-running space experiment has been on the astronauts themselves. As well as being the crew controlling amazing spacecraft, the astronauts have been guinea pigs for doctors on Earth who want to understand how the human body responds to an environment where gravity does not have an effect. On Earth, the body is built to fight gravity, with the heart pushing blood up to the head. This does not stop in the weightlessness of space, and astronauts get puffy faces and skinny legs as the blood displaces into the upper body. Similarly, the body is constantly refreshing the muscles and bones in response to the impact of gravity on Earth. In space this process stops, and bones thin, muscles weaken, and astronauts lengthen a few inches as their spines are no longer compressed. Exercise keeps the body healthy during stays in space of about a year. However, is there a time limit to space travel after which the body is too weak to survive back home?

Is suspended animation possible?

Many animals hibernate in winter. This is not just taking a long, deep sleep. Instead, the body processes slow right down to just enough to maintain survival, but not enough to support activity. If humans are to travel out of the solar system, the journeys will take many centuries. To survive, the crew is going to have to hibernate. In medical settings, patients have been chilled to low temperatures to curtail life processes during certain surgeries, and then warmed up and revived. It may be possible to do this for an indefinite period for space travel. Researchers have found they can use chemicals to slow metabolism in mice, and worms and fish have been placed in suspended animation, where all processes, including breathing, stop. They were then revived several hours later.

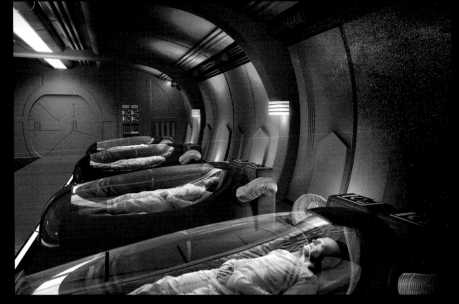

Is this the future of space travel?

Do we need biodiversity?

The great variety of life on Earth is what first drew our attention to it. Conservationists implore us to protect habitats and the multitude of species that live in them. But why? If the biosphere is a single system, it requires biomass, not biodiversity, to function. In other words, as long as there are enough trees, seaweeds, bacteria, and other minibeasts, the nutrients will be cycled through the system just fine. We don't need thousands of different types of tree, seaweed, and bacteria for that process to work. However, we do have thousands, even millions, of different types, and we cannot re-engineer the world's wildlife communities overnight. So we should protect what is there. Also, all that diversity offers solutions to problems in the future, such as diseases and climate change. Finally, the natural world is just amazing, so why not look after it?

Corals are among the most diverse habitats, and the most endangered.

The Great Biologists

GREAT BIOLOGISTS COME FROM MANY COUNTRIES, MANY TRADITIONS, AND ARE SPREAD THROUGH history, but they all have one thing in common: A wonder at the natural world. However, they have managed to turn this fascination with living things into a piece of knowledge, which reveals nature to be even more fascinating still. Let's take a look at the lives behind the discoveries that shaped biology.

Pliny the Elder

Born	23 CE
Place of birth	Novum Comum (Como), Italy
Died	August 25, 79 CE
Importance	Natural historian

Gaius Plinius Secundus was a career soldier who rose to the rank of cavalry commander before returning to semi-retirement in Rome. It was during this period that he produced his 37-volume *Historia Naturalis*. Toward the end of the reign of Nero, Pliny came out of retirement to become procurator in Spain, with responsibility for collecting taxes. After Nero's death in 68 CE, Rome was riven by civil war. It took a year of conflict before Vespasian, a friend of Pliny's, became emperor. Pliny became commander of the fleet in the Bay of Naples, a posting that led to his death when he went to investigate the eruption of Vesuvius.

Aristotle

Born	c.384 BCE
Place of birth	Stagira, Greece
Died	c.322 BCE
Importance	Founder of biology

Aristotle was an insatiable polymath; his thinking on such diverse topics as biology, physics, politics, agriculture, logic, and even dance shaped their development for centuries. He was the first to organize human knowledge into distinct disciplines, such as mathematics and biology. Of the 200  or more treatises Aristotle composed, only 31 survive, and those only as lecture notes and draft manuscripts. The quality of Aristotle's scientific research is remarkable. He accurately described features of insects that were not confirmed until the invention of the microscope.

William Harvey

Born	April 1, 1578
Place of Birth	Folkestone, England
Died	June 3, 1657
Importance	Discoverer of blood circulation

The son of a merchant, William Harvey was educated at Cambridge University and the University of Padua in Italy. Returning from Italy in 1602 to start work as a doctor, Harvey married Elizabeth Browne, the daughter of Queen Elizabeth I's physician. Harvey then became physician to James I, who succeeded Elizabeth, and to James's son Charles when he became king. Harvey maintained his close relationship with the royal family through the unrest of the English Civil War and was present at the Battle of Edgehill, the first major engagement of the conflict.

Jan Baptist van Helmont

Born	January 12, 1580
Place of birth	Brussels, Belgium
Died	December 30, 1644
Importance	Investigator of plant growth

Van Helmont studied philosophy and classics, dabbled in theology and law, and finally settled on medicine. He then threw away his books, declaring his education to be "senseless prattle." He honed his medical skills during a plague outbreak in Antwerp in 1605. He received several offers to become a private physician but turned them down. In 1609, van Helmont married into a noble family and retired. He published little until toward the end of his life. Shortly before his death, he gave his son responsibility for publishing his writings.

Nehemiah Grew

Born	September 26, 1641
Place of birth	Coventry, England
Died	March 25, 1712
Importance	Pioneer of botanical anatomy

Grew was the son of a clergyman who sided with the Parliamentary forces during the English Civil War. When Charles II was restored to the throne, Grew's father lost his income. Grew began a medical practice in Coventry but later moved to London. He was already showing an interest in plant anatomy, and in 1672 the Royal Society appointed him curator to the Society for the Anatomy of Plants. Grew was the first to successfully extract magnesium sulphate salts (Epsom salts) from the supposedly medicinal waters of the spa town of Epsom, Surrey.

Antonie van Leeuwenhoek

Born	October 24, 1632
Place of birth	Delft, Netherlands
Died	August 26, 1723
Importance	Inventor of microscope

Antonie van Leeuwenhoek was apprenticed to a linen draper. He received no higher education and spent most of his life in Delft, setting himself up in business as a fabric merchant. In 1676 he served as the trustee of the estate of the painter Jan Vermeer. It is probably the fact that he was untainted by the scientific dogma of the day, coupled of course with his skill in microscope making, that helped him make many discoveries in the history of biology, such as sperm cells, blood cells, and microorganisms. His findings began opening up the microscopic world to science.

Carl Linnaeus

Born	May 23, 1707
Place of birth	Råshult, Småland, Sweden
Died	January 10, 1778
Importance	Inventor of classification system

Carl Linnaeus's father was a keen gardener and passed on this love to his son. Carl studied medicine at the University of Uppsala, where he spent most of his time studying plants. (Doctors of the time prepared drugs derived from medicinal plants.) After a period in the Netherlands he took up a professorship at Uppsala in 1741. There, he restored the university's botanical garden, arranging the plants according to the system of classification he had devised. He inspired his students to travel the world collecting samples that swelled the pages of his *Systema Naturae* book.

Jan Ingenhousz

Born	December 8, 1730
Place of birth	Breda, Netherlands
Died	September 7, 1799
Importance	Discoverer of photosynthesis

As a doctor in London, Ingenhousz was a supporter of the practice of variolation, a form of inoculation against smallpox that involved using samples of the live virus. During an epidemic in Hertfordshire in 1767, he successfully immunized 700 people and went on to provide the same service to the family of King George III. He traveled to Vienna in 1768 to inoculate the family of the Austrian empress Maria Theresa and ended up spending several years there as court physician. Ingenhousz was also fascinated by electricity and invented an apparatus for generating static electricity.

Georges Cuvier

Born	August 23, 1769
Place of birth	Montbéliard (then Germany, now France)
Died	May 13, 1832
Importance	Discoverer of extinction

Georges Cuvier studied comparative anatomy in Stuttgart and worked as a tutor after graduating. He joined the new National Museum of Natural History in Paris in 1795, and soon won recognition as the world's leading expert on the anatomy of animals. He was reputed to be able to reconstruct the complete anatomy of a previously unknown fossil species from just a few fragments of bone. Cuvier was appointed to government positions by Napoleon, including inspector-general of public education, and continued as a state councilor under three successive kings after Napoleon's fall.

Jean-Baptiste Lamarck

Born	August 1, 1744
Place of birth	Bazentin-le-Petit, France
Died	December 18, 1829
Importance	Proposer of early theory of evolution

Today Lamarck is chiefly known as the author of a disproven theory of heredity. He was born into a family with a strong military tradition. He joined the French army, campaigning in Germany in 1761 where he demonstrated bravery under fire. In later life, as professor at the National Museum of Natural History in Paris, Lamarck almost single-handedly set up a new discipline, the study of invertebrates. But he never won the respect enjoyed by his contemporaries, such as Cuvier. He struggled with poverty throughout his life and when he died he was given a pauper's burial.

Alexander von Humboldt

Born	September 14, 1769
Place of birth	Berlin, Germany
Died	May 6, 1859
Importance	Explorer

The son of an army officer, Humboldt and his brother were raised by their mother after their father's death in 1779. After a disinterested stab at engineering, Humboldt discovered a passion for plants and geology. In 1796 he began to travel extensively, including making an epic trip to South America during which he explored the Orinoco River and set a world mountain climbing record on an ascent of Mount Chimborazo in the Andes. Humboldt won great fame—rivers, mountains, and towns all over the world were named in his honor.

Gideon Mantell

Born	February 3, 1790
Place of birth	Lewes, England
Died	November 10, 1852
Importance	Discoverer of dinosaurs

Physician, geologist, and paleontologist Gideon Mantell discovered four out of the five genera of dinosaurs known in his time. The son of a shoemaker, Mantell trained as a doctor. Returning to Lewes, he became a partner in a busy practice, sometimes attending to 50 patients a day. But he still found time to pursue  his passion for geology, often in the company of his wife, Mary Anne Woodhouse, whom he married in 1816. The couple separated in 1839. An accident in 1841 left Mantell with crippling spinal injuries and in 1852 he died from an overdose of opium he had been taking for the pain.

Charles Darwin

Born	February 12, 1809
Place of birth	Shrewsbury, England
Died	April 19, 1882
Importance	Proposer of theory of evolution

Charles Darwin was without doubt one of the most influential thinkers who ever lived—his theory of evolution by natural selection changed the way we look at the world. Darwin came from a scientific background; his father, Dr. R.W. Darwin, was a physician, and his grandfather, Dr. Erasmus Darwin, was a renowned  botanist. It was Darwin's five-year survey trip to South America aboard HMS *Beagle* from 1831 to 1836 that provided him with the impetus to develop his theory. He didn't publish until 1859 and was almost pipped to the post by Alfred Russel Wallace, who was having similar ideas.

Mary Anning

Born	May 21, 1799
Place of birth	Lyme Regis, England
Died	March 9, 1847
Importance	Fossil hunter

Mary Anning, described as "the greatest fossilist the world ever knew," was fortunate to have been born on Britain's Jurassic Coast—parts of Dorset and Devon—which is rich in fossils. According to family tradition, Anning, aged one,  survived a lightning strike that killed three others. When her father died in 1810 she sold fossils to make ends meet. As a woman in what was then very much a man's world she struggled to get the recognition she deserved for her finds, which included the first plesiosaur. She died of breast cancer in 1847, a few months after becoming an honorary member of the Geological Society of London.

Gregor Mendel

Born	July 22, 1822
Place of birth	Heinzendorf (now Czech Republic)
Died	January 6, 1884
Importance	Founder of Genetics

Austrian monk Gregor Mendel was born on the family farm. At school he showed an aptitude for learning and enrolled at the Philosophical Institute of the University of Olmütz, where he distinguished himself in physics and math, and tutored in his spare time to make ends meet. Against the wishes of his father, Mendel began studying to be a monk in 1843. It was at the St. Thomas Monastery in Brno that he would make his pioneering discoveries. Mendel eventually became abbot, which brought a halt to his research. When he died, his work was still largely unknown to the wider world.

Louis Pasteur

Born	December 27, 1822
Place of birth	Dole, France
Died	September 28, 1895
Importance	Discoverer of germ theory of disease

The son of a sergeant major, as a boy Pasteur had a talent for drawing and painting. He studied chemistry at the University of Strasbourg, where he made the discovery that certain molecules could exist as mirror images (or left- and right-handed versions) of each other. He invented pasteurization when he was asked to solve the problem of wine spoiling and is credited with saving the French wine industry. As a result of this, and for much else, when he died following a stroke he was hailed as a national hero in France. He was one of the founders of preventative medicine whose work saved countless lives.

Eugenius Warming

Born	November 3, 1841
Birthplace	Mandø, Denmark
Died	April 2, 1924
Importance	Plant ecologist

After a childhood in the Jutland countryside, Warming began a course of natural history at the University of Copenhagen in 1859. He then took a three-year break to act as an assistant on a fossil hunting expedition in Brazil, before returning  to complete his studies. Becoming the botany professor at the University of Copenhagen in 1882, Warming led several expeditions to Greenland, Venezuela, and the West Indies, creating a survey of vegetation in the temperate, tropical, and Arctic zones. This work was presented in *Oecology of Plants* (1895), an early contribution to ecology.

Alfred Russel Wallace

Born	January 8, 1823
Place of birth	Usk, Wales
Died	November 7, 1913
Importance	Proposer of theory of evolution

Naturalist Alfred Russel Wallace formulated his own theory of evolution by natural selection at the same time as Charles Darwin was working on his. Wallace was largely self-taught, his formal education limited to six years in a one-room school. He worked as a surveyor and was a keen amateur naturalist who read widely on the subject. Between 1854 and 1862 he collected thousands of plant and animal specimens in the Far East. It was there that he came up with his theory of evolution, which he shared with Darwin. In addition to his scientific pursuits, Wallace was also keenly interested in socialism and spiritualism.

Robert Koch

Born	December 11, 1843
Place of birth	Clausthal, Germany
Died	May 27, 1910
Importance	Pioneer of microbiology

The son of a mining engineer, at the age of five Koch apparently announced to his parents that he had taught himself to read by studying the newspapers. In 1862, he studied medicine at the University of Göttingen. He worked at the general hospital in Hamburg before going into  general practice and then becoming a district medical officer. His groundbreaking work on the anthrax bacillus was carried out in a home laboratory without proper scientific equipment in his four-room flat in Wollstein. Koch was awarded many prizes and medals, including in 1905 the Nobel Prize for Physiology or Medicine.

Ivan Pavlov

Born	September 26, 1849
Place of birth	Ryazan, Russia
Died	February 27, 1936
Importance	Investigator of animal learning

As the son of a priest, Ivan Pavlov attended a theological seminary, but abandoned theology in favor of science. Pavlov had an outstanding academic record and in 1890 he was invited to direct the Department of Physiology at the Institute of Experimental Medicine, a post he held for 45 years until the end of his life. It was here that he carried out his seminal work on the conditioned reflex. Following the October Revolution, Lenin issued a decree noting the "outstanding scientific services of Academician I.P. Pavlov, which are of enormous significance to the working class of the whole world."

Hans Krebs

Born	August 25, 1900
Place of birth	Hildesheim, Germany
Died	November 22, 1981
Importance	Discoverer of citric acid cycle

Hans Krebs is known for his remarkable work on the metabolic reactions of respiration in all living things. The son of a Jewish physician, Krebs was forced to leave his position at the University of Freiburg in 1933 and move to England to escape Nazi oppression. He continued his research at Cambridge University for the next two years before moving to the University of Sheffield and later  Oxford. Krebs was elected a Fellow of the Royal Society of London in 1947, won the Nobel Prize for Physiology and Medicine in 1953, and was knighted in 1958.

Martinus Beijerinck

Born	March 16, 1851
Place of birth	Amsterdam, Netherlands
Died	January 1, 1931
Importance	Discoverer of viruses and nitrogen cycle

Beijerinck is best known as a soil microbiologist and virologist and as one of the founders of environmental microbiology. He came from a poor family and was home-schooled by his father until he was 12. It was only due to the support of an uncle that he was able to study chemistry at Delft Polytechnic. He went on to obtain his doctorate from the University of Leiden in 1877. Beijerinck was plagued by bouts of depression throughout his life. He was also a perfectionist, which made him a very demanding, and unpopular, teacher. He never married, believing that marriage would interfere with his work.

Konrad Lorenz

Born	November 7, 1903
Place of birth	Vienna, Austria
Died	February 27, 1989
Importance	Founder of ethology

Lorenz had an interest in animals from an early age, keeping a variety of species, many of which he found on explorations around his home. He also helped to provide care for sick animals from the nearby Schönbrunn Zoo. He was a German army physician from 1942 to 1944, ending up as a prisoner of war in the Soviet Union.  Returning to Austria in 1948 he became head of the Institute of Comparative Ethology at Altenberg. In 1973, Lorenz, together with Karl von Frisch and Nikolaas Tinbergen, was awarded the Nobel Prize for Physiology or Medicine for their discoveries in animal behavior.

Francis Crick

Born	June 8, 1916
Place of birth	Northampton, England
Died	July 28, 2004
Importance	Co-discoverer of DNA structure

Francis Crick began his scientific career in physics, obtaining a BSc from University College London in 1937. During World War II he worked on the design of mines at the Admiralty Research Laboratory. In 1947 he made the decision to switch from physics to biology, joining the Medical Research Council. There he worked on the X-ray crystallography of proteins, obtaining his PhD in 1954. His friendship with James Watson, beginning in 1951, was a major influence on his career. In later years Crick devoted himself to research on how the brain works and the nature of consciousness.

Rosalind Franklin

Born	July 25, 1920
Place of birth	London, England
Died	April 16, 1958
Importance	Co-discoverer of DNA structure

Franklin was a pioneering molecular biologist whose researches contributed to the unlocking of the puzzle of DNA's structure. Her father apparently tried to discourage her ambition to be a scientist, preferring that she be a social worker instead. At the X-ray crystallography unit at King's College, London, she obtained the images that confirmed Watson and Crick's theories regarding DNA. Franklin died from ovarian cancer in 1958. In 1962, Watson, Crick, and Franklin's supervisor Maurice Wilkins won the Nobel Prize. The prize cannot be awarded posthumously.

James Lovelock

Born	July 26, 1919
Place of birth	Letchworth Garden City, England
Died	–
Importance	Proposer of Gaia Hypothesis

James Lovelock graduated as a chemist from Manchester University in 1941. He started work at the Medical Research Council in London but a scholarship took him to Harvard University Medical School in Boston in 1954. In 1961 he became professor of chemistry at Baylor University College of Medicine in Houston, Texas, and while there also worked on lunar and planetary research at the Jet Propulsion Laboratory. Since 1964 he has mainly worked independently, authoring 200 scientific papers and filing around 50 patents.

James Watson

Born	April 6, 1928
Place of birth	Chicago, USA
Died	–
Importance	Co-discoverer of DNA structure

At the age of 15, Watson transferred from high school to the University of Chicago as part of an experimental program for gifted children. Watson won a scholarship to the University of Indiana for postgraduate study, gaining his doctoral degree in zoology in 1950. He became fascinated by the DNA molecule and went to work on the problem with Francis Crick at the Cavendish Laboratory in Cambridge. Watson became director of the Cold Spring Harbor Laboratory, New York, in 1968. He was forced to resign in 2007 following accusations of racism.

E.O. Wilson

Born	June 10, 1929
Place of birth	Birmingham, Alabama, USA
Died	–
Importance	Sociobiologist and conservationist

Edward Wilson is the world's leading authority on myrmecology, the study of ants. He is also known for his work on sociobiology. Wilson studied biology at the University of Alabama and at Harvard University, where he obtained his doctorate in 1955. As a child Wilson suffered an eye injury that limited his depth perception. This, coupled with the onset of partial deafness during his adolescence, meant that Wilson was unable to continue his early studies in ornithology, which required keen eyesight and a good ear, and switched instead to entomology.

Lynn Margulis

Born	March 5, 1938
Place of birth	Chicago, USA
Died	November 22, 2011
Importance	Proposer of endosymbiosis theory

Lynn Margulis graduated from the University of Chicago in 1957, and soon after, married astronomer Carl Sagan. In 1960, Margulis earned a master's degree in zoology and genetics from the University of Wisconsin, followed in 1965 by a PhD in genetics from the University of California, Berkeley. She taught at Boston University from 1966 until 1988. Her theory of endosymbiosis was greeted with some skepticism, but it is now widely accepted. She is one of three American members of the Russian Academy of Natural Sciences.

Jane Goodall

Born	April 3, 1934
Place of birth	London, England
Died	–
Importance	Primatologist

Primatologist and ethologist Jane Goodall is renowned for her work on chimpanzee behavior. She left education aged 18, working as a movie production assistant before traveling to Africa where she became assistant to paleontologist and anthropologist Louis Leakey. Together with Leakey she began to study chimpanzees, and in June 1960 a camp was set up in the Gombe Stream Game Reserve from where she began her decades-long observations of chimpanzee behavior. She stayed in Gombe until 1975. In 1977 she founded the Jane Goodall Institute in California.

Richard Dawkins

Born	March 26, 1941
Place of birth	Nairobi, Kenya
Died	–
Importance	Evolutionary biologist

Richard Dawkins's childhood was spent in Kenya, where his father was stationed during World War II. The family returned to England in 1949. Dawkins achieved his master's and, in 1966, his doctorate degree in zoology at Oxford University, studying under Nobel prize-winning ethologist Nikolaas Tinbergen. In 1971, he published his first book, *The Selfish Gene*, which he described as an attempt to put right misunderstandings about Darwinism. In 1995 Dawkins was named the first Simonyi Professor of Public Understanding of Science at Oxford. He held the post until 2008.

BIBLIOGRAPHY AND OTHER RESOURCES

Books

Carey, Nessa. *The Epigenetics Revolution: How Modern Biology is Rewriting Our Understanding of Genetics, Disease, and Inheritance.* 2012.

Carroll, Sean B. *Endless Forms Most Beautiful: The New Science of Evo Devo and the Making of the Animal Kingdom.* 2005.

Carson, Rachel. *Silent Spring.* 1962.

Darwin, Charles. *On the Origin of Species.* 1859.

Dawkins, Richard. *The Selfish Gene.* 1976.

Diamond, Jared. *The Third Chimpanzee: On the Evolution and Future of the Human Animal.* 1991.

Gould, Stephen Jay. *Wonderful Life: The Burgess Shale and the Nature of History.* 1989.

Kratz, Rene Fester and Donna Rae Siegfried. *Biology Essentials For Dummies.* 2011.

Lane, Nick. *The Vital Question: Energy, Evolution, and the Origins of Complex Life.* 2015.

Mukherjee, Siddhartha. *The Gene: An Intimate History.* 2016.

Ridley, Matt. *Genome: The Autobiography of a Species in 23 Chapters.* 1999.

Schrödinger, Erwin. *What Is Life? The Physical Aspect of the Living Cell.* 1944.

Shubin, Neil. *Your Inner Fish: A Journey into the 3.5-Billion-Year History of the Human Body.* 2008.

Skloot, Rebecca. *The Immortal Life of Henrietta Lacks.* 2010.

Sterelny, Kim. *Dawkins vs. Gould: Survival of the Fittest.* 2001.

Thomas, Lewis. *The Lives of a Cell: Notes of a Biology Watcher*. 1974.

Watson, James D. *The Double Helix.* 1968.

Weiner, Jonathan. *The Beak of the Finch: A Story of Evolution in Our Time.* 1994.

Wilson, E. O. *Sociobiology: The New Synthesis.* 1975.

Yong, Ed. *I Contain Multitudes: The Microbes Within Us and a Grander View of Life.* 2016.

Museums and Places to Visit

Academy of Natural Sciences, Philadelphia, USA

American Museum of Natural History, New York City, USA

Australian Museum, Sydney, Australia

Beaty Biodiversity Museum, Vancouver, Canada

Beijing Museum of Natural History, Beijing, China

California Academy of Sciences, San Francisco, USA

Carnegie Museum of Natural History, Pittsburgh, USA

Deutsches Museum, Munich, Germany

Dinosaur National Monument, Colorado, USA

Field Museum of Natural History, Chicago, USA

Hungarian Natural History Museum, Budapest, Hungary

Indian Museum, Kolkata, India

Jurassic Coast Museums, Dorset and East Devon, UK

La Plata Museum, La Plata, Argentina

La Specola, Museum of Natural History, Florence, Italy

Lee Kong Chian Natural History Museum, Singapore

Micropia, Museum of Microbes, Amsterdam, The Netherlands

Muse, Science Museum, Trento, Italy

Museum of Natural History, Berlin, Germany

Museum of Natural Sciences, Brussels, Belgium

Nairobi National Museum, Nairobi, Kenya

National Museum, Rio de Janeiro, Brazil

National Museum of Natural History, Paris, France

National Museum of Natural Sciences, Madrid, Spain

National Museum of Nature and Science, Tokyo, Japan

National Science Museum, Pathum Thani, Thailand

Natural History Museum, London, UK

Natural History Museum, Vienna, Austria

Naturalis Biodiversity Center, Leiden, Netherlands

Olduvai Gorge Museum, Ngorongoro Conservation Area, Tanzania

Ontario Science Centre, Toronto, Canada

Origins Centre, Johannesburg, South Africa

PaleoWorld Research Foundation, Jordan, USA

Proyecto Dino, Lake Barreales Paleontological Center, Neuquén, Argentina

Science Center NEMO, Amsterdam, The Netherlands

Shanghai Natural History Museum, Shanghai, China

Smithsonian Institution, Washington, D.C., USA

State Darwin Museum, Moscow, Russia

Te Papa Museum, Wellington, New Zealand

Universeum, Gothenburg, Sweden

Wellcome Collection, London, UK

Wyoming Dinosaur Center, Thermopolis, USA

Zigong Dinosaur Museum, Zigong City, Sichuan, China

Archives and Preserved Equipment

Francis Crick archive, Wellcome Library, London, UK

Francis Crick personal archive, University of California at San Diego, USA

Darwin Papers, University of Cambridge, UK

Down House, Charles Darwin's home, Downe, Kent, UK

The Papers of Rosalind Franklin, Churchill Archives Centre, Churchill College, University of Cambridge, UK

Alexander von Humboldt papers, Berlin State Library, Germany

Jane Goodall Archive, Duke University, North Carolina, USA

Robert Koch Museum and Mausoleum, Robert Koch Institute, Berlin, Germany

Krebs Papers, University of Sheffield, UK

Antonie van Leeuwenhoek instruments and specimens, Royal Society, London, UK

Linnean Collections, Linnean Society, London, UK

Konrad Lorenz Archive, Konrad Lorenz Institute for Evolution and Cognition Research, Klosterneuburg, Austria

James Lovelock Archive, Science Museum, London, UK

Gideon Mantell papers, National Library of New Zealand, Wellington, New Zealand

Lynn Margulis papers, Library of Congress, Washington, D.C., USA

Mendel Monument, The Abbey of St. Thomas, Brno, Czech Republic

Louis Pasteur Museum, Tomb and Papers, Pasteur Institute, Paris, France

Alfred Russel Wallace Manuscripts, Natural History Museum of London Archives, UK

James D. Watson Collection, Cold Spring Harbor Laboratory Archives, NY, USA

Websites

Actionbioscience www.actionbioscience.org

Catalogue of Life www.catalogueoflife.org

Encyclopedia of Life www.eol.org

Encyclopedia of Life Support Systems greenplanet.eolss.net

Jane Goodall Institute www.janegoodall.org

Khan Academy www.khanacademy.org

The Nobel Prize www.nobelprize.org/educational

Nuffield Foundation www.nuffieldfoundation.org/practical-biology

Tree of Life Web Project www.tolweb.org/tree

Wikispecies www.species.wikimedia.org

Apps

Biology: Changing the world

Biology Tutoring Videos

Complete Biology

Encyclopedia of Life

Gene Index HD

Genetics and Evolution

OnScreen DNA Model

GLOSSARY

archaea
A group of primitive, single-celled organisms. Previously thought to be bacteria, genetic analysis of these organisms showed them to be distinct lifeforms, perhaps even more primitive than bacteria. Archaea are found today living in extreme environments, such as hot springs.

autotroph
An organism that does not consume food to survive. Instead, it taps into another source of energy to sustain its life processes. Nearly all plants are autotrophs, because they use the energy in sunlight to build glucose fuel by the process of photosynthesis.

biomass
A measure of living material used by ecologists to understand an ecosystem. An ecosystem can have a large biomass without necessarily having a large biodiversity, or variety of species. For example, a cornfield's biomass is largely made up of just one species: Corn.

centriole
An organelle, or internal structure in a cell, that is involved in the formation of the spindle during cell division. The spindle is composed of ultra fine tubes made from protein, which are attached to chromosomes and used to haul them around the cell.

chloroplast
An organelle that is the site of photosynthesis inside plant cells. The chloroplast is made up of stacks of membranes, upon which multiple chlorophyll molecules are arrayed. Chlorophyll absorbs energy from light, and reflects green light, making the cell, and plants in general, appear green.

chromatid
One half of a chromosome during cell division. Each chromatid in a pair is an exact copy of the other. They are connected at a single point called the centromere. In the final stages of cell division, the chromatids are pulled apart, making them into separate chromosomes.

cytoplasm
The liquid inside a cell, surrounding the nucleus and contained within the cell membrane. Cytoplasm is mostly water, with some other chemicals dissolved in it.

endoplasmic reticulum
A network of membranous tubes that is seen in most animal and plant cells. The endoplasmic reticulum (ER) is where the cell makes important chemicals. Smooth ER makes lipids, while rough ER makes proteins.

fruit
The botanical term "fruit" has a wider meaning than just culinary fruits. Botanically, fruit is the structure that develops around the seed of a flowering plant. After pollination, the seed develops in the ovule at the heart of the flower, and the rest of the ovary grows into a fruit. A fruit's primary role is to protect and disperse the seed. Some are meant to be eaten by animals, and the seed passes through the gut and is deposited far from the parent tree—with a dollop of fertilizer. Other fruits catch the wind or float in water.

genotype
A description of the two sets of genes inherited from the parents. Geneticists try to understand how a particular genotype gives rise to a set of measurable characteristics.

genus
The taxon, or classification group, above species. A genus contains a group of closely related species. For example, the *Panthera* genus contains the lion, tiger, leopard, and jaguar. The human genus is *Homo,* and our species is *sapiens.* All other members of the *Homo* genus are now extinct.

Golgi apparatus
A mass of membranes seen in animal and plant cells, especially those that secrete material such as resin or mucus. Materials that are to be released from the cell are parceled up into membranous spheres called vesicles by the Golgi apparatus. A vesicle's membrane merges with the cell membrane, opens up, and releases its contents to the outside of the cell.

heterotroph
An organism that must consume the bodies, waste, or remains of another organism to survive. All animals and fungi are heterotrophs. They all rely on plants to introduce energy into the food chain.

histone
A structural protein in chromosomes. DNA is wrapped around histones so it can be stored safely in tightly coiled units.

organelle
A structure inside a cell that is visible through a powerful microscope. Organelles are only seen in eukaryotic cells—the cells of organisms such as animals, plants, and protists. Bacteria and archaea do not have them. Each organelle performs a specific role in the cell. The large ones such as chloroplasts or mitochondria have their own membranes.

phenotype
The outward—and measurable—expression of an organism's genes.

protist
A single-celled organism that is made of cells that contain organelles. Amoebas and algae are examples of protists.

ribosome
A small organelle that is used to build proteins in the cell. Each protein is made by the ribosomes translating genetic information from a strand of RNA into a sequence of amino acids.

saprophyte
An organism that digests its food externally from the body. Fungi are the main saprophytes. They secrete digestive enzymes into their surroundings, and then absorb the simple nutrients released by the enzymes' action on any material that is there.

vacuole
A large storage sac seen inside plant cells.

vascular plant
A plant that has vessels running through its roots, stems, and leaves. These vessels transport water from the roots to the leaves, and sugars from the leaves to the rest of the plant. Ferns, conifers, and flowering plants are all examples of vascular plants, while mosses and seaweeds are non-vascular plants.

INDEX

Cataloging-in-Publication Data has been applied for and may be obtained from the Library of Congress.

ISBN 978-1-62795-093-0

Series Concept and Direction: Jeanette Limondjian
Design: Bradbury and Williams
Editor: Meredith MacArdle
Proofreader: Julia Adams
Picture Research: Clare Newman
Consultant: Dr. David H. A. Fitch
Cover Design: Igor Satanovsky

Publisher's Note: While every effort has been made to ensure that the information herein is complete and accurate, the publishers and authors make no representations or warranties either expressed or implied of any kind with respect to this book to the reader. Neither the authors nor the publisher shall be liable or responsible for any damage, loss, or expense of any kind arising out of information contained in this book. The thoughts or opinions expressed in this book represent the personal views of the authors and not necessarily those of the publisher. Further, the publisher takes no responsibility for third party websites or their content.

SHELTER HARBOR PRESS
603 West 115th Street Suite 163
New York, New York 10025

For sales in the U.S. and Canada, please contact
info@shelterharborpress.com

For sales in the UK and Europe, please contact
info@worthpress.co.uk

Printed and bound in China by Imago.

10 9 8 7 6 5 4 3 2 1

PICTURE CREDITS

BOOK

Alamy: 19th Era 38b, Alila Medical Images 57t, All Over Images 72tr, Nigel Cattlin 43r, Chromicle 32r, 50r, 86b, Tim Cuff 105t, 138tr, Dorling Kindersley Ltd 82tr, Michael Dwyer 139tl, Everett Historical Collection 138br, 139bl, Paul Glendell 120bl, Granger Historical Picture Archive 6b, 28r, 97b, 137bl, Interfoto 40b, 89b, 137br, Jacopin 112t, Keystone Pictures USA 87bl, 91b, 93b, 137tr, Megan Lewis / Reuters 98t, Maspix 106c, Molekuul. be 115t, Photo Researchers, inc. 25l, 92, 132tr, 133tl, Photos 12 102b, Pictorial Press Ltd 32l, 94t, 135bl, 135br; 138bl, Reuters 103b, Science Photo Library 42t, Martin Shields 113, Jeremy Sutton-Hibbert 112b, The Granger Collection 134tr, 135tr, The Natural History Museum 22tr, 49l, The Science Photo Company 7cr, 71br, World History Agency 14t, 65l, 136tr; **Getty Images:** Bettmann 58/59, Corbis Historical 12tr, Hulton Archive 114r, Nancy R. Schiff 139tr; **iStock:** Gnagel 12bl, Homo Cosmicos 11t; **IUCN:** Red List 62br; **Library of Congress:** 28l, 51tr; **Mary Evans Picture Library:** 11br, 16b, 17t, 44mr; **NASA:** 104t, 130t; **NOAA:** 89tr, 126t; **Roy Williams:** 8tr, 8cl, 8cmr, 8clm, 8br, 9tcl, 9tr, 9cml, 9blb, 9bcr, 104b, 124, 125; **Science Photo Library:** 98b, Max Alexander 139br, A.Barrington Brown, Wolfgang Baumeister 8bl, Biophoto Associates 3, Gonville and Caius College 94b, Henning Dalhoff 131t, Dennis Kunkel Micropscopy 69cl, 87tr, Reinhard Dirscherl 120cr, Gunilla Elam 114l, Equinox Graphics 116, Sam Falk 76bl, Gary Hincks 79, Dr Charles Mazel, Visuals Unlimited, Inc. 103t, D. van Ravenswaay 106/107b, Science Science 76bl, Victor Habbick Visions 117r, Visuals Unlimited, Inc./Carole & Mike Werner 83; **Science & Society Picture Library:** 30l; **Shutterstock:** 3D Stock 117l, Alila Medical Media 84l, Alinabel 81tr, Artartty 8bc, Arturasker 73brc, Ba dins 41t, George W. Bailey 49br, Naza Basirun 59r, Stephane Bidouze 118cr, Bildagentur Zoonar GmbH 124, Blue Ring Media 35tr, BMP 73brb, Rich Carey 126b, Calmara 95t, Jose Luis Calvo 62l, Choksawatdikorn 69br, Cynoclub 9blt, Andrea Danti 55tr, Designua 67l, 77bl, Elle Picgrafica 97t, Dirk Ercken 7t, Geertweggen 120cl, Gio.tto 7cl, 109, Gorosan 107l, Pawel Graczyk 26l, Jurin Grooverider 124, Jubal Harshaw 34t, H Helene 14c, Ibreakstock 121cl, Igor ZD 22l, Jonathan C Photography 118b, Joyfull 50bl, Kaliva 90, Kamomeen 125, Ox Karol 124, Kateryna Kon 73bl, 121tl, Sebastian Kaulitzki 121cr, Peter Kotoff 73brt, Brian Kinney 110b, Lebendkulteren.de 8tc, 66b, Wichawan Lowroongroj 77bc, Lynea 33r, Mahey 37b, Anukool Manston 73tr, Don Mammoser 119cl, Mimohe 22bc, Mopic 122t, Morphart Creation 78tl. 82b, 84r, Nechaevkon 78bl, Nobeastsofierce 69tr, Hein Nouwens 78tr, Evan Novostro 130b, Pakpoom Nunjui 66l, Dicky Oesin 9bcl, Olesandrum 9cl, Onemu 129br, Ozja 46b, Heidi Paves 119tr, Maryna Pleshkun 9trc, 124, Alex Polo 125, Alexander Potapov 121b, Randall Reed 121tr, Michal Rosskothen 49r, Rusn123 72b, Dennis Saba 124, Guy J Sago 125, Menno Schaefer 22bl, Schantz 61bl, Science Photo 108r, Selivanar 111b, Ashray Shah 91t, Sipa Photo 127b, Snap Galleria 29br, Snow Cake 68b, Kippy Spilker 35l, Stan de Haas Photography 60r, Aleksey Stemmer 9cmr, Super Prin 119tl, Talvi 9br, The Learning Photographer 48l, Rathiya Thongdumhyu 120tc, 123br, Irina Tischenko 125, 131b, Tomatilo 119cl, TT Photo 95b, 125, Marian Uradnik 67r, Usagi_p 70b, Vdaix 81b, Mike Versprill 63b, Dennis van de Water 110t, Bogan Wankowicz 119br, Kev Williams 38tr, Jolanta Wojcicka 9tl, Vladimir Wrangel 124, Wire_man 54r, Ziviani 53br; **Thinkstock:** Abelstock.com 68t, Andrew Ags 128t, Georgios Art 134br, Ikon Studio 86tr, Patrice Latron/Eueliosl Look at Sciences 115b, Yang Mingqi 128b, Molekuul.be 129l, Ramdan Naim 7b, 53l, Tonis Pan 127tr, Photos.com 15b, 39ml, Daniel Prudek 64l, ttsz 122b, 122-123t; **U.S.National Library of Medicine:** 12/13, 24l, 37t, 40l; **USGS:** 63t; **Wellcome Library, London:** 6tr, 16tr, 18b, 21tr, 23l, 25r, 26r, 27t, 27b, 29l, 29tr, 33tl, 36br, 39tr, 41b, 42b, 45t, 45b, 46c, 48r, 52r, 56l, 58tr, 59t, 61tl, 65tr, 65br, 66r, 132bl, 132br, 133tr, 133br, 134tl, 134bl, 135tr, 136tl, 136tr, 136br, 137tl, 138tl: **Wikipedia:** Atynoise 101, Pedro Brisda 99, Graeme Churchard 107t, Matdir 105b, Tim Vickers 111t, Vossman 96b, 2l, 2r, 4, 5, 6cl, 10, 10/11, 13b, 14b, 15t, 16l, 17b, , 18t, 19t, 19b, 20tr, 20ml, 20mr, 20bl,

21ml, 21mr, 21bl, 23r, 24t, 30r, 31, 33bl, 34l, 34br, 35br, 36t, 43l, 44tl, 44b, 46r, 47t, 47bl, 50tl, 51b, 52l, 53mr, 54l, 55l, 55br, 56r, 57b, 58l, 60l, 61r, 62tr, 64r, 70tl, 71tl, 74t, 74b, 75bl, 77tr, 80, 88, 96t, 100, 102t, 108b, 133bl.

TIMELINE

Alamy: A. F. Archive, Chronicle, Jason Bye, David Norton Photography, Everett Collection Historical, Granger Collection, Interfoto, Lebrecht Authors, M. Ramirez, MEPL, North Wind Picture Archive, Photo Researchers, Inc., Prisma Archivo, Trinity Mirror, World History Archive; **Library of Congress: Mary Evans Picture Library: NASA: Shutterstock:** 360b, Alessia Pierdomonico, Brett Anderson, Hung Chung Chih, Juan Ci, Daxiao Productions, drserg, Everett Historical, Igor Golovniov, Diego Grandi, Amy Nichole Harris, Georgios Kollidas, Kullanart, Morphart Creation, Nicku, Noko Nomad, Sean Pavone, Reidl, Fedov Selivanov, Travel Light, Vvoronov, Yankane, Zhu Difeng; **Science Photo Library:** A. Barrington Brown/Gonville & Caius College; **Thinkstock:** Demerzel21, Feel Like, Georgios Art, Velikova Oksana, Papadimitriou, Photos.com; **Wellcome Library, London; Wikipedia.**

REVERSE TIMELINE

Science Photo Library: Biophoto Associates, Eye of Science, Steve Gschmeissner, Power and Syred, Vshyukova

Publisher's Note: Every effort has been made to trace copyright holders and seek permission to use illustrative material. The publishers wish to apologize for any inadvertent errors or omissions and would be glad to rectify these in future editions.

CONTRIBUTORS

Richard Beatty is an Edinburgh-based writer, editor, and lexicographer, specializing in biology and the history of science. As well as authoring numerous books and articles, he has worked for many years on the Oxford English Dictionary, researching the history of "scientific" and other words. **Pages 32–51**

Leon Gray studied zoology at University College London. Since graduating, Leon has written and edited more than 100 nonfiction books, mainly about science, technology and the natural world. He lives in Aberdeenshire, Scotland, with his wife, three children, and two lurchers. **Pages** 54–55, 60–61, 82–91

Dr Jen Green is a full-time writer with over 350 books in print. She specializes in books and websites on nature, environment, science, geography and history. As a Royal Literary Fund Consultant Fellow she regularly gives writing skills workshops to university students. **Pages** 58–59, 62–64, 72, 76–80

Tim Harris has written dozens of books for children and young people on science and natural history. **Pages** 52–53, 56–57, 66–71, 92–117

Tom Jackson, editor, is a science writer based in the United Kingdom. Tom specializes in recasting science and technology into lively historical narratives. After almost 20 years of writing, Tom has uncovered a wealth of stories that help create new ways to enjoy learning about science. **Pages** 6–11, 65, 73–75, 81, 118–131

Robert Snedden has worked for over thirty years as an editor and writer. He is an insatiable autodidact in the grand Scottish tradition and delights in learning new things. **Pages** 12–31 and 132–139